THE CURE WITHIN

ALSO BY ANNE HARRINGTON

The Dalai Lama at MIT (editor)

The Placebo Effect: An Interdisciplinary Exploration (editor)

Reenchanted Science: Holism in German Culture from Wilhelm II to Hitler

Medicine, Mind, and the Double Brain

THE CURE WITHIN

A History of Mind-Body Medicine

ANNE HARRINGTON

W. W. Norton & Company New York • London

Printed in the United States of America
First Edition

Manufacturing by RR Donnelley, Harrisonburg
Book design by Chris Welch
Production manager: Andrew Marasia

Library of Congress Cataloging-in-Publication Data

Harrington, Anne, 1960–
The cure within : a history of mind-body medicine / Anne Harrington.
p. ; cm.
Includes bibliographical references and index.
ISBN 978-0-393-06563-3 (hardcover)
1. Medicine, Psychosomatic—History. 2. Mind and body—History. I. Title.
[DNLM: 1. Psychosomatic Medicine—history. 2. Mental Healing. 3. Mind-Body Relations
(Metaphysics) 4. Psychophysiologic Disorders. WM 11.1 H299c 2008]
RC49.H333 2008
616.00199—dc22
2007030906

W. W. Norton & Company, Inc., 500 Fifth Avenue, New York, N.Y. 10110
www.wwnorton.com

W. W. Norton & Company Ltd., Castle House, 75/76 Wells Street, London W1T 3QT

3 4 5 6 7 8 9 0

To John: my toughest critic and best friend

CONTENTS

Acknowledgments

This book has been long in the making, and over the years it has been enriched by contact with numerous people and institutions. I found my original inspiration for this book during a six-year stint as a core member of the MacArthur Foundation Network on Mind-Body Interactions. I am grateful for all the opportunities afforded me through that association, and for all the stretching conversations I had over the years with the members of that group: Robert M. Rose (director of the network), Richard J. Davidson, J. Allan Hobson, Martha K. McClintock, John T. Cacioppo, Stephen M. Kosslyn, John F. Sheridan, David M. Spiegel, Esther May Sternberg, and Eve Van Cauter.

I am grateful to Harvard University's Interfaculty Mind, Brain, Behavior Initiative for supporting me in my early interest in the placebo effect and my efforts to catalyze constructive dialogue across the biomedical sciences and the social sciences on the topic. I also owe a debt to the Mind and Life Institute, and especially its chair and CEO, Adam Engle, for the opportunity to participate in projects and dialogues concerned with relations between Buddhist philosophy and practice, science, and health. Thanks also to the Shanghai Qigong Research Institute in Shanghai, China—and especially to my primary mentor, Dr. Chu—for first opening up the world of medical qigong practice for me in immediate experiential ways. Early on, I had the good fortune to collaborate with colleagues at the National Library of Medicine on an exhibition entitled "Emotions and the Brain" that gave an early outing to some of the ideas in this book; my appreciation here especially to Esther Sternberg, Elizabeth Fee, and my fellow guest curator on the exhibition, Theodore Brown. Finally, from 2003 to 2004, I enjoyed a sabbatical year at the Bristol Institute for Advanced Studies, in the University of Bristol, UK, where I sketched the first draft of the ideas of this book. My thanks here go to Karine Taylor (administrator) and to the then-provost of the institute, Bernard Silverman.

For interviews and various personal communications, my thanks go to Herbert Benson, Lisa Berkman, Richard Davidson, Marceil Delacy, Howard Fields, Daniel Goleman, Jon Kabat-Zinn, Feng Lida, Robert Rose, Arthur Shapiro, David Spiegel, Francisco Varela, and Peter Hahn (who also shared rare books from his private library at the University of Heidelberg with me).

For comments on chapters of the manuscript at various stages, I thank Laura Harrington, Jon Kabat-Zinn, Stephen Kosslyn, Katinka Matson, Daniel Moerman, Charles E. Rosenberg, Sam Schweber, and Ilina Singh. For research assistance at various stages of this project, I am grateful to Ziad Obermeyer, Juliane Oepen, Ilina Singh, and Kevin Stoller.

In the final leg of this project, Kevin Stoller, an advanced graduate student at Harvard University in the Department of the History of Science, also proved himself a wonderfully perspicacious reader and editor. I feel very fortunate to have had the benefit of his intelligent eye and many suggestions.

Various arguments in this book were first aired and discussed in lectures given at the National Center for Complementary and Alternative Medicine at the National Institutes of Health; the Department for the History and Philosophy of Science at Cambridge University; the Program in History of Medicine at Yale University; the Program in Science, Technology, and Society at MIT; the Eidgenössische Technische Hochschule in Zurich; the Department of Sociology at the University of Bristol; and the Bios Centre at the London School of Economics. In addition, over the past several years, I undertook to test-drive much of the material of this book in an undergraduate course I offered in the Faculty of Arts and Sciences at Harvard University called "Stories under the Skin: The Mind-Body Connection in Modern Medicine." The students in that class were my first real audience, and I owe more than I can probably properly acknowledge to their responses, challenges, questions, and insights.

Real thanks also to my agent, Katinka Matson, for always believing in the project, and for providing encouragement at several crucial points. My unfailingly kind and insightful editor at W. W. Norton, Angela von der Lippe, saw the trade-book potential in a manuscript written in an overly academic style and pushed me to let a bit more fresh air and sunshine into the prose; I am grateful for her challenge. Copyeditor Elizabeth Pierson was a joy with whom to work. Elly Truitt did a splendid job setting up and implementing a system for finding and securing permission to publish all the images in this book. And in the last months before final delivery of the manuscript, Abby Hudson, my assistant at Harvard, stepped into multiple breaches: helping with the input of corrections, with copyright permissions, with the scanning of images, and much else besides.

My most personal thanks go to my husband, John Durant, who supported me through this whole process in more ways than I can list, read more versions of chapters than he should have, and always told me the truth about how he thought I was doing. This book is dedicated to him.

The Cure Within

Introduction

STORIES, SCIENCE, AND CULTURE UNDER THE SKIN

Why do we fall ill? Most of the time, most of us believe we know the answer. Some illnesses are due to infection from a virus or bacterium; some to the breakdown of proper functioning of some organ in our body; some to a deficiency or excess of some crucial hormone, enzyme, or other chemical we need to be healthy. We believe too—more and more—that genes often play a role in illness, that certain diseases run in families. At the same time, we willingly accept that some diseases are triggered or exacerbated by a toxic environment (living by a chemical factory), a bad diet (too much saturated fat), or unwise habits (cigarette smoking).

All these understandings have in common one thing: a belief that physical symptoms of illness have physical causes. If the physical causes of our illness are not immediately obvious to the unaided senses, then people expect their doctors to use X rays, ultrasound, CT scans, laboratory analysis of blood or tissue, or surgery to look more deeply for the cause of what ails. Once the doctor can "see" what is wrong, the hope is that he or she will be able to tell the patient how to fix him- or herself: what drug to take, what kind of surgery to have, what change in diet or lifestyle to make. For many these days, this way of thinking about illness includes even that class of disorders we still call "mental illness."

Quite often, this physicalist way of thinking about illness works. Patients take the antibiotic and recover from their infection, learn to inject themselves with insulin and normalize their blood sugar levels, have surgery and learn that their cancer has gone into remission, or take the antidepressant and find they can get out of bed again in the morning.

What happens, though, when one or another of our standard physicalist approaches to illness does not work, fails to satisfy either doctors or patients? For example, what happens when someone develops all sorts of somatic symptoms, and is clearly suffering, but medical tests fail to turn up anything physically wrong? In 2003, the *Journal of the American Medical Association* (*JAMA*) published an essay called "Patients Like Linda." Linda suffered from chronic symptoms suggestive of a degenerative neuromuscular disorder. Nevertheless, "numerous competent doctors, including two neurologists, found her examination normal, and myriad negative results from laboratory tests and imaging studies supported their conclusion that no identifiable organic illness was causing her myalgia, fatigue, joint pain, and insomnia."

What all the doctors suspected was that Linda was suffering more from problems of life than from problems of the body, but Linda did not want to be told that, and became angry when doctors talked gently about her seeing a psychiatrist. Steven Ringel, the author of the 2003 *JAMA* essay, mused on the meaning of Linda's defensiveness; her clarity that, if her suffering was not strictly physical in origin, it was also not strictly real. She was not wrong to be defensive. In fact, patients like Linda, lacking clear physical markers of disease, are generally not taken seriously by their doctors:

> Over the years I've learned that almost every aspect of our health care system is more responsive to the needs of patients with major organ failure. We miss the mark for those who fear they have a serious yet undiagnosed disease and who have unexplained pain, weakness, fatigue, headaches, mood changes, or interrupted sleep. . . . To the extent that they cannot function, they are truly disabled. To

> make matters worse, physicians don't want to take care of these people, and insurance companies don't want to pay for their care.[1]

So doctors become impatient and insurance companies skeptical when confronted with "patients like Linda," who make no sense, conceptually, to physicalist medicine. Meanwhile, the legions of Lindas in our society continue to suffer. Do any genuinely effective or empowering options exist for them?

The shortcomings of physicalist medicine, however, do not end there. Sometimes a patient can present with a disease that all agree is real, but the patient is then informed that little or nothing can be done about it. Perhaps there are no therapies; or perhaps such therapies as exist are unreliable, unsafe, or merely palliative rather than curative. For many patients, especially those who have learned to bring a consumerist sensibility to their medical care, situations like this can produce a sense of helplessness, even betrayal. Some patients may choose to give up, but others choose to get mad. What happens then? What do they do? To what or to whom do they turn?

Conceptual shortcomings and therapeutic shortcomings—these would seem to be serious enough perceived deficiencies in the edifice of physicalist modern medicine. But there is one other that still bears noting: an existential deficiency. The physicalist approach to illness falls short, especially for patients, because it denies the relevance of the kinds of questions people so often ask when they become ill: Why me? Why now? What next? "Your illness has no meaning," patients are told instead. "You just had bad luck. You inherited a genetic vulnerability, or you got an infection." Bad stuff happens. Don't try to make sense of it all.

And yet, throughout history, people have always tried to make personal sense of the suffering of illness. Historically, Western cultures have provided people with a stockpile of religious, moral, and social stories to help them answer the great "why" questions of their suffering, and to connect their experiences to some larger understanding of their identities and destinies. Traditionally, some illness was said to be a consequence of personal sin, other illness was an outcome of evil in the community, or a test of faith, or a message from God to change one's life. But now? To the extent that modern physicalist medicine does offer any kind of story about illness, it is a story that is as impersonal as they come: it is all about the disease rather than the patient, and it is articulated using a specialized vocabulary of tissue, blood, and biochemistry.[2]

When one can take a shot of antibiotics and feel better in the morning, an impersonal story like that may feel fine. But when medicine lets one down in other respects—by failing fully to validate the complexity of one's suffering, for example, or by being unable to deliver a cure—then physicalist talk about tissue, blood, and biochemistry may begin to feel hollow indeed. At this point, some people begin to long for something better—a better story. In his 1995 book *The Wounded Storyteller*, sociologist Arthur Frank talks eloquently about this need. "Stories," he reminds us, "repair the damage that illness has done to the ill person's sense of where she is in life and where she may be going."[3] At the same time, Frank goes on, many people no longer feel able to draw comfort or meaning from the traditional sources for such stories—family histories, folk beliefs, religious traditions—because such traditional sources no longer have credibility in the modern world.

This may be true, but is there anything more to say? Here I come finally to my point. In fact, there is a third way of thinking about physical illness—and of telling stories about physical illness—that is neither the traditional way that has lost credibility, nor the standard physicalist way that lacks existential relevance. This third way also claims to be rooted in science. Indeed, it often claims to have a scientific understanding of illness that is more complete than that provided by the physicalist stories of mainstream medicine. It insists, though, that there is more to physical illness than can be seen just in the body; and more to healing than can be found in just pills and shots. Mind matters too: how one thinks, how one feels, what kind of personality or character one has or cultivates. For stories of this third kind, questions like "Why me? Why now? What next?" are not meaningless after all, but exactly the right questions—and for medical and scientific reasons, not just moral and existential ones. And this being so, it follows that there may be other ways than those of physicalist medicine by which to heal the body of the real disorders that ail it.

What do the stories that put mind and personality in the picture look like? We all know them. In one, a business executive with a high-pressure job and a terrible temper resists the advice of friends and family to calm down and is finally felled by a massive heart attack. In another, a man suffers from a chronic disorder that only flares up at socially inconvenient times, leading his wife to conclude that he is being manipulative or acting out some kind of repressed anger. And there is the one about the cancer patient who, refusing to believe the gloomy prognosis of her doctors, inspires everyone with her good cheer and determination to live, and then in fact experiences a remarkable recovery.

Where do these stories of mind-body illness and mind-body healing come from? What is their relationship both to modern medicine on the one side and traditional religious and folk stories about illness on the other? What can their prominence in our culture today teach us about our own imperfectly secularized experiences of illness, the curious ways in which we navigate a path between science and sense-making in our own time? What can they teach us too about discontents and countercurrents within the profession of mainstream medicine itself, the ways that doctors and scientists also feel and respond to the shortcomings of the physicalist story of illness? This book takes an historical approach to tackling these kinds of questions. It does so, moreover, in a way that puts the spotlight, first and foremost, not just on the science, not just on the practices, not just on the institutions, but on the stories of mind-body medicine: where they have come from, what they have helped generate over time, what functions they serve in our society today.

Put another way, this is a book that seeks to claim mind-body medicine for cultural history—to show how it functions as a far-flung and omnivorous discourse that does not respect the boundaries we try to set up between the professional and the popular, but that in different ways shapes the talk, work, and experiences of all of us alike, be we doctors, scientists, patients, health gurus, paperback writers, journalists, or web-bloggers. In attempting a cultural history of mind-body medicine, I focus on stories because I believe this is the best way to carve the subject up at its joints. In mind-body medicine we are confronted not with an integrated vision or program but with a patchwork of approaches and understandings that pull in many different directions. Some approaches emphasize our power to heal ourselves. Others (particularly those highlighted by critics) emphasize our vulnerability to the psychological influences of others. Some are nostalgic for traditional forms of wisdom.

Others want to lead us into a new age of high-tech scientific investigation. Some paint a picture of the mind in colors infused with a lush neoromantic sensibility. Others portray it in the angular lines of mechanistic modernism.

Each of these stories, I insist, has its own history, comes from somewhere, and can be told. Yes, there are interactions, blended approaches, borrowings, and appropriations, but our historical preoccupations with stress are not the same as our efforts to harness the power of positive thinking, and these are different again from our commitment to the confession of trauma as a healing act. For these reasons, as I turn my attention in this book to each new class of story within mind-body medicine, I invite my readers to embark with me on a fresh historical exploration.

At this point, some readers may be wondering when I am going to mention the seventeenth-century French philosopher René Descartes. After all, everyone knows, don't they, that it was Descartes who, with his notorious philosophy of mind-body dualism, launched medicine on the path toward its current unsatisfying vision of human mind-body relations? Don't all the stories of mind-body medicine have in common a commitment to repairing the wrong he did to us more than three centuries ago? "Blame it on Descartes," a journalist reporting in 1997 on mind-body research activities at the University of Rochester insisted:

> One morning, so the story goes, the 17th-century philosopher/scientist awoke from a dream, inspired to pursue the study of metaphysics. The culmination of this study was his conclusion that body and mind were separate entities. At that point he and the Roman Catholic Church struck a deal: As a man of science, Descartes would restrict himself to the study of human anatomy, leaving the mind and soul to the care of the church.[4]

In fact, I have found little evidence that the history of mind-body medicine is best seen as beginning in a moment of original philosophical sin, which is then followed by a series of brave efforts to set things right again. The varied champions of mind-body healing explored in this book rarely, if ever, directly engage with the work of Descartes; and when they do hold him up as the enemy of their cause, they tend consistently to caricature and misrepresent him. Descartes certainly contributed broadly to modern philosophical sensibilities and understandings of mind and body. His importance to the history of mind-body medicine, however, is less as a real philosopher and more as a symbol of modern errors, a foil against which modern champions of one or another story of mind-body integration express their nostalgia for a fantasized premodern past when we all were whole and integrated, mind and body.[5]

A theoretical interlude

I have been using the word "story" freely in my remarks so far, but what do I mean by this word? "Story" is, of course, part of common speech, where it often has a suspect status, certainly to those people who value their science. It can refer to an informal account that lacks rigor ("what's the story?"), an account that is personal and therefore probably biased ("that's his story"), or worst of all, an account that is frankly untrue ("stop telling stories!"). When I speak of stories in this book, I do not mean to imply any of those common understandings. Stories can be more or less true or untrue, more or less grounded in scientific data, and more or less self-interested. What distinguishes a story from other kinds of communication is a certain kind of structure that follows three basic rules:

1. A story selects among a pool of potentially relevant events and sequences them—normally chronologically—in ways that direct the imagination along a particular trajectory ("After Cinderella's own mother died, her father married a wicked woman with three daughters of her own").
2. A story identifies the relationship among these sequenced events, often but not always by declaring or implying a thread of linear causality running across them ("The prince saw how beautiful Cinderella was, and in that moment he fell in love with her").
3. A story ends, often with a final event that implicitly affirms the importance of certain values. Certain actions lead to good outcomes ("Cinderella married the prince"), and other actions lead to bad ones

("Cinderella's stepmother and her ugly daughters lived the rest of their days under a cloud of shame and disgrace"). The best stories leave us with a sense of satisfaction for this reason.

Good stories, however, are not just talk. Many of the best ones are also scripts, or guides to action: things that provide us with a store of specific, if flexible, social roles, cues, and reference points that tell us how to behave and even—sometimes—how to feel (what to experience) in different situations.[6] The stories of mind-body medicine are at their most powerful when they are functioning as scripts like this, inspiring people to turn their backs on conventional remedies, defer to new kinds of healers, embark on unaccustomed travels, undertake new kinds of practices, and more. To paraphrase (and slightly mangle) a famous comment by the sociologist Erving Goffman, who brilliantly elucidated the dramaturgical side of everyday life: all the world of mind-body medicine is not, of course, a stage, but the crucial ways in which it isn't are not easy to specify.[7]

Sometimes, of course, a particular chosen script fails, and the most disappointing or destabilizing failures of all come when the body itself fails to perform its appointed role. This patient's ulcers turn out to respond better to antibiotics than to psychotherapy or stress-reduction exercises; the cancer patient who maintained an inspiringly positive attitude is unable to rally and dies. Bodies cannot be made to act every part that may be chosen for them in any number of different stories; they may be flexible, but they are not infinitely plastic.

At the same time, bodies also participate in history, even if historians do not always know how to deal well with that fact. I first became aware of this years ago, when doing research on the history of hypnosis as part of some work for my dissertation. I remember sitting in the Bibliothèque Nationale in Paris, books and old medical journals piled up on the table before me, and thinking, "I have no idea what to do, because it is clear that the understandings of hypnosis are not just changing over time. The mental and physiological *experience* of hypnosis—what it *is*—is changing too; and changing in ways that clearly reflect changing social expectations and mores." Eighteenth-century patients of Anton Mesmer felt animal magnetism shooting through their bodies; patients of J. Chastenet de Puységur no longer did, but instead gave evidence of having access to heightened, even supernormal mental abilities. By the second half of the nineteenth-century, that had all stopped, and instead hypnosis had become a quasi-pathological phenomenon, with specific physiological profiles: catalepsy, lethargy, somnambulism. None of my teachers had ever suggested to me that bodies might have culture too, that life inside our skins might have a history; but there was the evidence in the hypnosis literature at least, clear as could be.

The larger implications of this insight are worth spelling out in a little more detail. It is well known that human beings in general respond to the expectations of people around them, including and especially the expectations of the authority figures or experts. This can introduce uncertainty into the efforts of social scientists trying to study various social and psychological phenomena. Economists know, for example, that when predictions of supply and demand become public knowledge, markets may falsify the predictions because investors and companies respond to the predictions—so-called Goodhart's law.[8] A range of other terms have been used to variously describe the challenges that come from studying human beings who respond to the fact that expert claims are being made about them, or that they are being studied: Hawthorne effects, experimenter expectancy effects, Pygmalion effects, subject-expectancy effects, and most broadly of all, looping effects.[9]

But are such effects seen not only in behavior but also in bodily experience? My experience in a library in France suggested they are, and later I found a handful of anthropologists and historians whose work on the relationship between folk belief, cultural values, and bodily experience further encouraged me. In his studies of the Kalahari bush people in the 1970s, for example, Richard Katz described the intensely painful "boiling energy" that rises up from the bellies of native healers during ritual healing dances, an experience that makes full sense in the context of their own cosmology but that has no obvious parallel in any experiences most people have in Western cultures.[10] In her comparative work on aging in Japan and North America, Margaret Lock found that when older women in Japan stop menstruating, they rarely experience the hot flashes and night sweats that are associated with menopause in North America. In fact, it turned out the Japanese have no word for a hot flash. Lock's view is that these vasomotor differences have their origins in the fact that female aging is not experienced as a sign of diminished status and worth in Japan in the same way as it is in North America.[11]

Also suggestive to me in this context was the work of historians of the body like Caroline Walker Bynum and Barbara Duden. In different ways, these scholars pointed to records of the past that are filled with reports of human beings having bodily experiences that differ significantly from anything in our experience today. People seem to have menstruated, lactated, experienced changes in skin and facial appearance, bled or failed to bleed in ways that make no sense to us—but that did make sense to people at the time. Do we just say that these reports from the past are deceptions, delusions, tales that were never meant to be taken literally? Or is there something potentially more interesting to say about such material? Perhaps reports like these are hints of the different ways in which people long ago tended to experience and even visibly display authoritative understandings of what bodies are and how they may legitimately work.[12]

Let me be clear. I do not think, as an historian, that we can just invent or "make up" our bodies anyway we want. Nevertheless, I have written this book against a background belief that the body is a genuinely mindful entity.[13] Therefore its experiences are likely to change over time when the "rules" of what counts as acceptable or possible experience change. Since, in modern times, such rules are increasingly set by medical and scientific proclamations, we should expect bodies today to be "listening" above all to scientists (or people they believe to speak with the authority of science) or to their doctors. We might have expected that more researchers of mind-body medicine, of all people, would have long since appreciated this possibility and have reflected on its larger implications, but few seem to have done so.[14] Perhaps one effect of this book will be to encourage more thinking along these lines.

Having taken some pains to clarify what I mean by stories and the roles they play in history, inside and outside the body, I want to make a further distinction between stories and something I will call "narratives." Stories are living, local, and specific. They are the things we read in books and newspapers, hear on the bus, tell over dinner, and use to guide behavior and experience. They refer to immediate, concrete events, people, scientific findings, and more. Narratives, however, are templates: they provide us with tropes and plotlines that help us understand the larger import of specific stories we hear, read, or see in action. They also help us construct specific stories of our own—including ones about our own experience—that others can recognize and affirm. We learn these narrative templates from our culture, not in the way we might formally learn the rules of grammar in school, but in the way we might unconsciously learn the rules of grammar at home—by being exposed to multiple individual examples of living stories that rely on them.

An example may be helpful. In 1997, Diana, Princess of Wales, was killed with her companion Dodi Fayed in an automobile crash in Paris as they attempted to flee the press. Almost immediately, stories about the event began to proliferate in the tabloids and in internet discussions. A great many of these stories drew on the plotlines and tropes of a narrative we might call "true love thwarted." Diana was the "people's princess" who had finally, after many years of unhappiness, found true love with "her knight in shining armor."[15] An unfeeling and envious world did not understand and resented their happiness, and in the end destroyed them both. In literature, the most familiar rendering of this particular narrative of tragic love is probably Shakespeare's *Romeo and Juliet.* It is telling in this context that various admirers of Diana and Dodi quickly took to calling them the "Romeo and Juliet of the 20th century," the "star-crossed lovers of the 90s," etc.[16]

The stories of mind-body medicine depend on a small set of narrative templates like "true love thwarted." None of these narratives have titles in the real world, but for the purposes of this book I have invented a title for each: "The Power of Suggestion," "The Body That Speaks," "The Power of Positive Thinking," "Broken by Modern Life," "Healing Ties," and "Eastward Journeys." I have settled on these six narratives after close to a decade of firsthand observation, participation, and reading in the world of American mind-body medicine. I am persuaded that they are real, not in some archetypal or platonic sense,[17] but in the way that the abstract rules of etiquette for different cultures are real and, with time and patience, can be identified and named with a fair degree of certainty, even if they are nowhere written down. Like the rules of etiquette, the effects of these narratives are everywhere in the ways that people talk, think, and behave; they help make sense of much that would otherwise seem mysterious or arbitrary about the business at hand; and "natives" invariably recognize a description of a narrative—they know it without realizing it—once it is pointed out to them.

Our modern rules of etiquette came from somewhere, and their histories can be told—not directly, but through the evidence of changes in sensibility and behaviors documented in laws, letters, programs for schooling, imaginative literature, and more.[18] It is the same with the narratives of modern mind-body medicine.

Previews

To see how this works in practice, let us look briefly at what is to come. Chapter one, "The Power of Suggestion," explores the historical emergence of what I consider to be the *skeptical or debunking* narrative of mind-body medicine. People under the influence of charismatic authority figures can have bodily experiences that look like solid biology but that turn out, on closer inspection, to really be "all in the head." Let the buyer beware.

Some of the fundamental ingredients for this narrative lie deep in the past. Once, much of the Western world functioned comfortably with a narrative that said illness is often due to demonic possession, and cure is to be sought in such cases by exorcisms undertaken by a representative of God, a consecrated priest. To understand the emergence of suggestion, we must first bear witness to the partial secularization of the demonic possession narrative, as it was transformed into Mesmer's ritual of animal magnetism, and then follow the debunking of Mesmer's cures as products not of fraud but of the imagination—a capacity of the mind that can lead to illusory experiences. By the late nineteenth century, at the high point of a new medical fascination with hypnosis, we see the pieces come together to create a new, modern narrative about something called "suggestion": a psychological force that is still wrought by powerful authority figures but that produces experiences more illusory than real.

Chapter two, "The Body That Speaks," is the *detective* narrative of mind-body medicine. Historically, its deep roots lie in understandings of the healing power of religious confessional rituals; but it emerges in its specifics in the late nineteenth century as a challenge to those who dismissed the disorder of hysteria and all its symptoms as a product of mere suggestion. "The body that speaks" insists instead that some of the experiences of ailing bodies, while they may not make any organic sense, still need to be taken seriously: the symptoms are a coded message that can be deciphered and brought into the open. When the message is decoded, healing becomes possible.

The relevant history here begins in the world of early Viennese psychoanalytic practice, then moves to the battlefields of World War I and the veterans' hospitals of World War II. Our story also takes us into a comparative exploration of German-speaking and American psychosomatic medicine. The former was distinctly romantic, alternative, buoyed along by larger holistic cultural strains of the time, and then increasingly politicized as Germany succumbed to Nazism in the 1930s. The latter was funded by well-heeled foundations and sought alliances with physiologists and a place in the medical mainstream.

After the collapse of the American movement in the 1970s, stories featuring ideas of the body as an entity with coded secrets to share did not vanish from the cultural scene; instead they found new life as resources for personal empowerment in various alternative feminist and holistic medical movements on the margins of American mainstream medicine. The message derived from such stories now was that a patient needs to trust her body and what it might be saying to her, whatever skeptical or unsympathetic doctors might say.

From this narrative about a body made ill by the mind's secrets, we move to "The Power of Positive Thinking" in chapter three, which is all about a mind possessed of great powers to heal the body of its ailments. This is the *secular miracle* narrative of mind-body medicine. Its twin roots lie in French medical efforts to make sense of the alleged supernatural healings happening at the Catholic shrine of Lourdes and in populist American efforts to realize the Gospel promise of healing through faith in immediate practical terms. Only in the 1970s, in the heat of the first holistic medicine movement, do we see a partial blending of the medical and populist versions of this narrative: Norman Cousins laughed his way back to health, and thousands of physicians saluted him. The placebo effect, previously vilified as a product of mere suggestion, was made over into the prime medical mover of positive thinking's power and declared the new faith cure of our time.

Chapter four, "Broken by Modern Life," takes up the narrative of *lament* of mind-body medicine. This narrative tells us that we suffer from disease because we live in a modern world that challenges our energies beyond their capacities. "Stress"—a new concept that emerged from a marriage between interwar laboratory science and attempts to make sense of the breakdown of soldiers during World War II—is both the concept and the experience at the heart of this story. Postwar anxieties in America about the cost of prosperity on our emotional and physical well-being helped catalyze further uses of this narrative, especially through the growing focus on a tragic (yet, paradoxically, rather admired) victim of modern capitalist society: the Type A personality. The greatest risk for such personalities was heart disease, it was believed, and the effects of stress on the heart remained a focus of research into the 1970s. At the same time, these years also saw the rise of a new frontline response to the stress epidemic. Biofeedback emerged as a new laboratory-based technology of self-regulation that promised ways of undoing systematically what modern life had wrought.

In the 1980s, a new and terrifying disease emerged on the scene: AIDS. Its appearance coincided closely with the rise of new research that seemed to suggest that the nervous system had ways of communicating with the immune system and that therefore the immune system (not just the heart) might be affected by stress. Did stress, then, make AIDS worse? If so, could its impact be reduced? What was the effect of stress on the immune systems of other vulnerable groups: elderly people, Gulf War veterans, cancer patients? In this period, it seemed as if there was no place in the body where stress did not stalk, and stories of stress became about as dark as they had ever been.

From "Broken by Modern Life," we move to two alternative kinds of *redemption* narrative templates that both begin, in different ways, by emphasizing the effects of modern life on our health, but then offer two very different kinds of solutions to making things better. Chapter five, "Healing Ties," is the first of these. It is the *nostalgia* narrative of mind-body medicine. It insists that we suffer so much from stress, not because modern life is so overwhelming in its demands, but because it has robbed us of community and of intimacy, leaving us with no friends, no network of supportive comrades, to buffer and aid us in facing life's challenges. A return to community, an embrace of something people began calling "social support," is what we need to heal. Close-knit communities in heartland America where people do not die of heart attacks, children in orphanages who fail to grow properly, women with breast cancer who seem to live longer because they meet weekly to share their troubles: all these things and more buttress the plausibility of this narrative and help it to realize new forms of expression and uses for itself.

Chapter six, "Eastward Journeys," is the *exoticism* narrative of mind-body medicine. It also aims to help us find ways to improve our health and well-being, but it seeks the source of healing not so much in the caring communities we have lost but in the healing practices of ancient Eastern cultures we have never known. The deep roots of this narrative lie in long-standing Orientalist tendencies in Western culture to conceive the East as a foil to our own Western values and lifestyles, generally to our (i.e., the West's) advantage. For "Eastward journeys" to emerge in mind-body medicine, however, two things had to happen: first, the moral logic mediating comparisons of East and West had to be inverted—the East now had to become the spiritual exemplar for the world from which the West had much to learn; and second, the moral logic had to be blended with a health message—the practices and understandings of the East were not only morally but also medically exemplary.

This chapter explores three contrapuntal and distinct moments in this process, the historical emergence of three variants employing the basic "Eastward journeys" template in recent American mind-body medicine: the medicalization of meditation, especially transcendental meditation, in the 1970s; the American embrace of the Chinese practice of qigong in the early 1990s; and the rise, at the turn of the millennium, of a new research effort concerned with the health benefits of Buddhism, and in particular with what certain extraordinary people—adepts, lamas, monks—might be able to teach the rest of us about human potential.

These, then, are the six narrative templates that, from a cultural perspective, can help us make sense of the complex patchwork world of American mind-body medicine. Armed with a knowledge of these narratives, we are able to understand how mind-body medicine can be "about" so many different things at once, and how it can have an effect at the same time on so many different levels of our culture: the spiritual and the secular, the elite and the popular, the mainstream and the marginal. Any history purporting to shine light not just on the science but on the culture—the many *meanings*—of mind-body medicine cannot afford to leave out any of these levels.

Chapter One

THE POWER OF SUGGESTION

Suggestion: Presentation of an idea. That which is suggested; an intimation; hint. An evil incitement; temptation. The entrance into the mind of an idea or intimation, originated by some external fact or word, which tends to produce an automatic response or reaction, as hypnotic suggestion. "Many 'miracles' of healing, and of 'stigmatization,' become credible when verified in modern experience and explained by 'suggestion.' " (James Ward)

—Webster's New International Dictionary of the English Language, *1930 edition*

Mr. Wright was a very sick man. He had been admitted to the hospital with a diagnosis of lymphosarcoma, cancer of the lymph nodes. Tumors, some the size of oranges, infested his neck, groin, and armpits. He was on oxygen and sedatives, and when his cancer ceased to respond to any conventional therapies, his doctors gave up all hope of a recovery. He had not given up hope himself, however. This was because he had learned that the hospital in which he was staying had been chosen as an evaluation site for a new experimental drug (derived from horse serum) called Krebiozen, and he was persuaded that Krebiozen would be his miracle cure. He begged his doctor to give him some. Although he met none of the criteria for inclusion in the trial (patients had to have at least a three-month life expectancy), he was so persistent that his doctor finally relented; yes, he could have an injection. It was duly administered on a Friday, and this is how his doctor described the scene that greeted him on Monday morning.

> I had left him febrile, gasping for air, completely bedridden. Now, here he was, walking around the ward, chatting happily with the

> nurses, and spreading his message of good cheer to any who would listen. Immediately I hastened to see the others who had received their first injection at the same time. No change, or change for the worse was noted. Only in Mr. Wright was there brilliant improvement. The tumor masses had melted like snow balls on a hot stove, and in only these few days, they were half their original size!

Mr. Wright continued on this stunning recovery course until conflicting reports about the actual efficacy of Krebiozen appeared in the newspaper. His confidence undermined, Mr. Wright relapsed. At this stage, his physician resolved that the medical interest of the case justified a bit of duplicity. Mr. Wright was told he should not believe the newspapers: his relapse was due to the fact that the original injection of medication he had received had decayed and was therefore substandard. The hospital, however, was being sent a fresh and extrapotent new batch in a few days, and when it came, Mr. Wright would be one of the first to receive it. Hearing this, Mr. Wright now waited in a fever pitch of impatience until finally his physician announced that the new batch of medicine had arrived. A syringe was brandished, and Mr. Wright was injected—but not with another dose of Krebiozen. All he received was a shot of distilled water.

Nevertheless, Mr. Wright's response to this second injection was more dramatic than before. Tumors again receded, and before long he was declared a picture of health and sent home. (In fact, this man, who a week earlier could not breathe without extra oxygen, flew home in his own airplane.) For a time, all continued to go well. Then an American Medical Association announcement appeared in the press definitively denouncing Krebiozen as a worthless drug. Mr. Wright read the report, relapsed again, was readmitted to the hospital, and was dead in two days.[1]

The story of Mr. Wright was first reported in a psychiatric journal in 1957 by Rorschach test pioneer Bruno Klopfer, who was fascinated by the personality profile of this patient.[2] Since then, it has been frequently retold, sometimes rather breathlessly, especially in the alternative-medicine cancer literature. Skeptics may or may not believe it or may scoff that it is just one case and in itself proves nothing. But virtually everyone who tells or hears it in our own time understands what *kind* of story it is: it is a story about the double-edged power of suggestion.

What, though, do we mean by this? What is "suggestion," and what kind of power do we think it possesses? To understand, we need to see the way in which the story of Mr. Wright is grounded in the assumptions and tropes typical of the oldest and most generative narrative of mind-body medicine. This is a skeptical narrative that I call "the power of suggestion."

In its classic form, this narrative invariably begins by introducing at least two characters: a vulnerable, naïve, or needy person (often a patient, quite often a woman) and an authority figure (typically a doctor, healer, hypnotist, or priest, and invariably a man) who is believed to possess either personal charisma, special skills, powerful medicines, or expert knowledge that brooks no skepticism. If he says something will happen, it will! The narrative then classically recounts a series of exchanges, often quite ritualized, between these characters. The sequence and substance of these exchanges can and does vary, but in all of the iterations the patient is helpless to resist: he or she believes whatever is said, does whatever is said, and—strangest of all—physically experiences whatever is said. The authority figure's words and acts seem to open up channels of communication between the patient's mind and body that are normally impassable. He or she may shake with cold or perspire with heat, or may undergo surgical procedures without discomfort. Sometimes—as happened in the story of Mr. Wright—if he or she is ill, all of his or her symptoms may disappear, either on the spot or over time.

But usually—as also happened with Mr. Wright—recovery is only temporary. The narrative's conclusion is clear: suggestion's cures are at best palliative and at worst fool's gold. The patient under the influence of suggestion has not actually been cured by her doctor, but has instead brought on all the changes herself, using her own mind—and these changes remain, in a sense, only mental. Her body is less cured than it is tricked by both her mind and her doctor, in a collusion that bypasses the patient's own consciousness. Put another way, the powers of suggestion depend on a Faustian bargain in which the patient yields her autonomy to an external authority, lays her troubles at his feet, and hopes to receive in return access to powers and experiences she could otherwise never hope to enjoy. It is a sweet, almost erotic vision—but a dangerous one. For the effects of suggestion are ultimately illusory, and therefore in practice they are highly unstable. Unmask the reality behind the theater, undermine the patient's trust, and everything collapses in a moment.

Among the six narrative templates of modern mind-body medicine to be explored in this book, "the power of suggestion" is the one most likely to make us uncomfortable. It is true that we are sometimes intrigued by the idea that we possess deep and generally untapped powers to influence our bodily functioning, but we are also highly ambivalent about the idea that these powers are for others, and not us, to command. Yes, some stories about suggestion celebrate the uniquely trusting relationship between doctor and patient that enables suggestion to work its effects, but more common versions are scornful of patients who are susceptible to the suggestive interventions of authority figures. There is often an implication that, if suggestion can cure ailments, then these ailments must have been "all in the patient's head" in the first place. We may also be suspicious of the ethics of any doctor who uses suggestion as a therapeutic method for illness. At best, we may feel such a method undermines patients' autonomy; at worst, it puts them at risk of being stooges of the unscrupulous.

As a narrative with all of these connotations, "the power of suggestion" has existed in our culture for only about a century. Many of the key elements operating to give the narrative its real power are, however, much older. Indeed, we can only make full sense of the haunted, uneasy nature of "the power of suggestion" by appreciating its Janus-faced roots. On the one side, it draws on key elements within the centuries-old and now largely defunct narrative of demonic possession; on the other side, it draws just as heavily on the skepticism that has dogged possession and its successive secular analogues—first mesmerism, and then hypnosis—since at least the sixteenth century. Were these strange states of mind the product of powerful external forces—satanic, physical, or psychological? Alternatively, were they the product of outright fraud? Or again, were they perhaps the result of unwitting self-deception? No other narrative of modern mind-body medicine is as fundamentally conflicted about its own epistemological and ethical message as this one, and its history is the primary reason.

Possession, exorcism—and their first skeptics

To see how this all came to be, we need to begin with a phenomenon that has been documented in societies all over the world: possession. In the 1960s, anthropologist Erika Bourguignon analyzed available records on 488 different societies. Using just the data in the published literature, she found that 360 (74 percent) of these societies told possession stories and that 52 percent had ceremonies for treating cases of possession. If the available data had been more comprehensive, she argued, the percentages might well also have been higher.[3] Her data and similar work have led some anthropologists to suggest that human beings have some basic psychobiological machinery that enables, and perhaps even predisposes them to, experiences of possession.[4]

At the same time, there are no raw or unmediated possession experiences: every culture shapes its experiences of possession in accordance with its own beliefs, institutions, and social structure. The Judeo-Christian culture's encounter with possession goes back at least to the Judaic Second Temple Period (beginning around 200 B.C.), from when we have records of beliefs in the existence of roving demonic forces capable of entering the bodies of errant or unlucky folk.[5] When they did so, people no longer behaved normally, but engaged in wild, immoral, or improper conduct that endangered themselves and those around them. Often, they would also become physically ill—vomiting, breaking out in skin rashes, suffering spells of suffocation, and succumbing to violent convulsions.

What could be done for these unfortunates? Jewish tradition provided a range of incantations and other rituals for driving out demons. Later, of course, stories would be told of Jesus casting out devils in his role as itinerant healer. This familiar story from Mark is typical:

> And when He came to the disciples, He saw a great multitude around them, and scribes disputing with them. Immediately, when they saw Him, all the people were greatly amazed, and running to Him, greeted Him. And He asked the scribes, "What are you discussing with them?" Then one of the crowd answered and said, "Teacher, I brought You my son, who has a mute spirit. And wherever it seizes

> him, it throws him down; he foams at the mouth, gnashes his teeth, and becomes rigid. So I spoke to Your disciples, that they should cast it out, but they could not."
>
> He answered him and said, "O faithless generation, how long shall I be with you? How long shall I bear with you? Bring him to Me." Then they brought him to Him. And when he saw Him, immediately the spirit convulsed him, and he fell on the ground and wallowed, foaming at the mouth.
>
> So He asked his father, "How long has this been happening to him?" And he said, "From childhood. And often he has thrown him both into the fire and into the water to destroy him. But if You can do anything, have compassion on us and help us." Jesus said to him, "If you can believe, all things are possible to him who believes." Immediately the father of the child cried out and said with tears, "Lord, I believe; help my unbelief!"
>
> When Jesus saw that the people came running together, He rebuked the unclean spirit, saying to it, "Deaf and dumb spirit, I command you, come out of him and enter him no more!" Then the spirit cried out, convulsed him greatly, and came out of him. And he became as one dead, so that many said, "He is dead." But Jesus took him by the hand and lifted him up, and he arose.[6]

Here are all the key elements in a drama that would be played out regularly during the early history of Christianity: the individual made violent by demonic forces; the plea for help; the moral injunction ("if you can believe . . ."); the imperative command; the wrenching moment of the demon's departure; the crowd bearing witness. Tellingly, the word that came to be used for the rituals modeled on this and other similar New Testament stories of demon-possession was "exorcism," from the Greek word *exorkizein*, which means "to bind by an oath." In other words, exorcism was to be understood as a process by which an individual compels demons to depart by calling on the binding power of a higher authority—in the case of Mark's biblical story, the ultimate authority of God Himself acting through Jesus.[7]

The insistence that only a priest or consecrated man has the authority to exorcize in the name of God came much later in the history of Christian exorcism. It wasn't until the late sixteenth century that the Catholic

Church began to emphasize this plot element, in response to Protestants and dissenters who began suggesting that individual prayer and fasting might actually be sufficient to purge the devil from a possessed person. In this sense, the Church's new hard-line insistence that priests alone were qualified to wrestle with Satan was part of its larger campaign to reassert its authority against the Reformation.[8] In 1614, Pope Paul V issued a famous edict, the Rituale Romanum. This asserted the reality of demonic possession and outlined the procedures for exorcism by priests. Rather remarkably, these procedures remained on the books essentially unaltered until 1999.[9]

The role of the priest was precisely delineated in this new orthodoxy: he must function as a sort of spiritual physician-cum-surgeon, first confirming the presence of the demonic, then using his spiritual authority to expunge it. A considerable number of prescribed props and rituals conspired to make the priest's authority visible. He was required to don a surplice and purple stole and to utter certain prayers for protection (Litanies of the Saints, the Lord's Prayer, Psalm 54). He was then enjoined to call on the demon or evil spirit to make itself known. At this point, it was common for the suffering person to collapse in convulsions. These were assumed to be at once an admission of demonic presence and an acknowledgment of the priest's power to force the evil spirit to reveal itself. Indeed, it was understood that the demon must respond to the commands of the priest because demons were members of a universal hierarchy within which they were subordinate to God, and priests in their roles as exorcists were acting directly on behalf of God.[10] The priest might go on to interrogate the demon about its motives for possessing this person in particular; and again, it was believed that the demon was compelled to respond truthfully. The climax of the exorcism came when the priest used ritual words and gestures—making the sign of the cross, invoking the name of Christ—to banish the demon from the person's body. And usually, the demon went. When it did, the person's soul returned to reanimate his or her vacated body, typically with amnesia for all that had transpired.

Even as this highly elaborate narrative was still in process of formalization, however, the conditions for its destabilization were already being fostered. In the highly charged political atmosphere of the Reformation, when virtually all aspects of Catholic authority and practice were under scrutiny or attack, people began to raise the specter of fraud. Accepting in principle the possibility of demonic possession, some commentators suggested that not every case of apparent possession should be taken at face value. In the late sixteenth century, the prominent case of one Marthe Brossier in France would prove particularly significant in raising the profile of this new skeptical stance.

Marthe Brossier was a twenty-two-year-old woman living in a village in the French province of Berry when she began showing the classic signs of possession—wild behavior, convulsions, and so on. But some authorities believed that Marthe's symptoms of possession were more apparent than real. The girl was being manipulated, they said, by her family and the Catholic clergy. The family, they said, saw an opportunity for financial gain, while the clergy were interested in stirring up anti-Protestant sentiment. In 1598, Henry IV asked the physician Michel Marescot to lead an investigation of Marthe's case. What Marescot and his colleagues did, however, was remarkable—and, as it may well seem to us today, remarkably modern: they offered Marthe consecrated water—something that demons cannot abide—but lied and told her that it was in fact plain water. She had no reaction to it. They then read her passages from the *Aeneid* but lied again and told her that they were from the Bible. This time, Marthe responded with convulsions and every sign of distress. The conclusion from these and similar tests seemed clear, and Marescot summed it up with the memorable line "Nothing from the devil, much counterfeit, a little from disease."[11]

This is a striking moment in the history of exorcism's encounter with medicine. From now on, it would no longer be enough (if indeed it ever really had been) for a person simply to act in odd ways or to report powerful and disturbing experiences. He or she must also provide externally verifiable evidence that the experiences in question were demonically inspired. Marescot and his colleagues proposed the following as standard benchmarks: being able to speak and understand languages of which the possessed person had no prior knowledge; being able to discern secrets and predict future events; demonstrating abnormal strength and insensitivity to pain; and consistently demonstrating revulsion at holy things (contact with holy water, reading of Scripture, etc.).[12] In this world, demons were real; but so too were deception and delusion. Faced with the need to distinguish reliably among these things, rational people needed to have a battery of tests at their disposal.

From invisible demons to invisible fluids

But what if demons were not real? By the second half of the seventeenth century, the Church was facing a new kind of skepticism: skepticism not just about the credibility of particular cases of possession, but about the very possibility of possession itself. There were many sources for such general skepticism, including new natural philosophies that emphasized the powers of nature to produce effects previously attributed to the work of spirits, as well as new forms of anticlericalism according to which demon possession was simply a device used by priests to keep ignorant people subservient to their authority. As the political philosopher Thomas Hobbes complained in his classic work *Leviathan*, "By their demonology, and the use of exorcism, and other things appertaining thereto, [the priests] keep, or think they keep, the people in awe of their power."[13]

Appreciating the interweaving religious, philosophical, and political stakes here is important because it can help us make sense of an episode whose significance we might otherwise be likely to misinterpret: the showdown in the 1770s between the German exorcist Father Johann Joseph Gassner and the Viennese physician Anton Mesmer.[14]

Gassner was an exorcist whose ability to cast out devils was legendary. People came from all over to be healed, and in dramatic public performances—witnessed by crowds from all sectors of society—Gassner would oblige. Official records were made; competent witnesses testified to the extraordinary happenings. All agreed on the basic facts. On being presented with a supplicant, Gassner would typically wave a crucifix over his or her body and demand in Latin that, if the disease he was seeing had a "preternatural" source, this fact must be made manifest. The patient would then typically collapse into convulsions, and Gassner would proceed to exorcise the offending spirit.

Sometimes he added flourishes to this basic routine: in one dramatic instance, for example, he ordered the demon inside a woman to increase the poor victim's heartbeat and then to slow it down. Following the sec-

Father Johann Joseph Gassner (1727–1779) performing an exorcism on a young woman before a crowd of witnesses. Castelli, in Louis Figuier, *Les Mystères de la Science,* ca. 1770. *© Mary Evans Picture Library*

ond command, a witnessing physician was invited to examine the patient and declared her dead—he could find no pulse, he exclaimed; her heart had stopped! But Gassner remained calm and demanded that the demon responsible for these acts depart from the body of this woman at once. The command given, the woman stirred and rose to her feet before the crowd, alive and well.[15] No one could exorcise like Gassner.[16]

But the increasingly secular sensibilities of the time combined with the growing hostility of the civic authorities toward the Church to make Gassner's very successes the cause of his undoing. The medical profession complained; the local authorities complained; and in 1774, by order of Prince Max Joseph of Bavaria, a commission was set up to investigate all these goings-on. One of the experts invited to assist in the investigation was a young physician named Franz Anton Mesmer.

Mesmer was asked to assist because he seemed to have a perspective on Gassner's performance that would be very useful for the skeptics and others who wanted to rein in Gassner. Today, Mesmer is remembered—if he's remembered at all—as a charlatan, or a showman, or maybe as someone who discovered the existence of psychological processes that he did not himself properly understand. He considered himself, however, to be a child of the dawning enlightened, scientific age. He was interested in the larger implications of Newton's ideas about physical forces and gravitation; and he was skeptical, both of the old supernatural ideas about the world, and of the old authority structures of the Church.

Mesmer was particularly interested in the medical implications of Newton's theory of gravitation. Newton had suggested that the human body might contain an invisible fluid that responds to planetary gravitation, like the tides of the ocean. Taking up this idea, Mesmer performed an initial series of experiments in which he moved mineral

Franz Anton Mesmer (1734–1815).
© *Bettmann/Corbis*

magnets around the bodies of his patients. In response to such treatments, Mesmer's patients reported experiencing strong sensations of energy moving through their bodies. They also experienced all sorts of involuntary movements, including often violent convulsions. These convulsions left many patients feeling much improved, or even cured of their ailments. Apparently, mineral magnets could have great therapeutic value through their ability to influence human magnetic fluids in this fashion.

But then Mesmer made a further discovery: the mineral magnets were not actually necessary to the effects he could produce! By simply waving his hands over a patient's body, Mesmer was able to produce precisely the same results as he had before using mineral magnets. From this startling observation, Mesmer concluded that he himself had actually been the source of the invisible magnetic energy or force that had been benefiting his patients all along. He called this force "animal magnetism," and suggested that it had worked its therapeutic effects by rechanneling or refortifying the weakened animal magnetism in his patients.

More than anything else, it was Mesmer's new theory of animal magnetism that interested the Commission investigating Gassner's work, for they could see that it gave Mesmer a new way in which to think about the Christian drama of demonic possession and exorcism. In 1775, Mesmer gave several demonstrations in which he showed that he could first evoke convulsions in people and then dispel them with a peremptory gesture, just like Gassner. Such things, he said, were not what they seemed—they had nothing to do with the devil or the supernatural. They were, he said, simply the consequence of his manipulation of the wholly natural force of animal magnetism.[17]

Not spirits but an invisible magnetic force! It was a conclusion that helped give the authorities the rationale they needed to put a stop to Gassner's public career. From then on, the priest was forbidden to engage in public exorcisms. He was confined to a small parish and only allowed to perform private exorcisms under supervision. Gassner's problem was not that he had failed to help people, but rather that the narrative of exorcism he used to frame his therapeutic work no longer suited an increasingly secular, civic-minded age. As historian Henri Ellenberger wryly observed, "it is not enough to cure the sick; you have to cure them with methods accepted by the community."[18]

For our purposes, the end of Gassner's public career as an exorcist is important for how it paved the way for the rise of a new healing script. Once, when people fell into strange states of consciousness, collapsed into convulsions, and then submitted unwillingly but inexorably to the commands of a superior authority, they had been caught up in a struggle between evil and good, between a demon and one of God's official representatives on earth. Now such people were in the thrall of powerful physical forces that were just beginning to be discovered in the new Newtonian universe. And where once sufferers had submitted to the healing hands and words of priests, now increasingly they were invited to submit to the healing hands and words of physicians.

Nevertheless, the rise of animal magnetism was not a moment of mere disenchantment or demystification, in the way we sometimes imagine happens when science steps in to explain things that had previously been seen as having purely theological significance. Instead, the rise of animal magnetism signaled the emergence of a new kind of story and ritual that still followed the structure of the old exorcists' rites and still had the power to evoke a sense of awe and wonder.

Thus, the convulsions that had been so central to the drama of exorcism remained the central experience of a mesmeric healing; but now they were a sign not of a demon over which a priest had asserted authority but of the detection of a disease over which the magnetizer exercised control. Mesmer did not wear a priest's purple stole, but he did favor something hardly less imposing: a sweeping lilac taffeta robe that, according to some, sported esoteric celestial symbols. Again, he didn't use the traditional emblems of the Church's authority—Bibles, holy water, crucifixes—but he did employ a range of physical props that reflected the rising authority of the new natural philosophies of the eighteenth century. There were specially designed baths filled with water and iron filings, into which projected large iron rods; there were hanging mirrors that were supposed to reflect back and intensify the invisible energy circulating in the room; and there was mysteriously unfamiliar music playing in the background behind a purple curtain. The music, produced by a new invention called the armonica, was not just there to help create a mood; like the iron filings and mirrors, it was supposed to act as a physical aid to the intensification and conduction of the animal magnetism. Even on its own, the armonica was said to be capable of causing dramatic sensations in some patients.

The stage having been set, a typical treatment session by Mesmer or one of his assistants went as follows. First, patients looked deep into Mesmer's commanding eyes (the eyes were said to be particularly potent ports of entry and exit for the magnetic force). Then Mesmer made a series of sweeping gestures to direct magnetic fluid to appropriate parts of the patients' bodies. Typically, patients would begin to tremble and twitch before finally succumbing to the convulsive "crisis," a signal that Mesmer's personal animal magnetism had disrupted the unhealthy flow of animal magnetism in the patients' bodies. Finally, assistants helped the convulsing patients into a special recovery room where, after a while, they usually declared themselves to be feeling much better than before.[19]

Mesmer's therapeutic approach would later be the subject of much parody, and a certain historical memory of those parodies lingers in our own thinking; somehow, we "know" that his claims about animal magnetism were absurd.[20] At the time, though, there was nothing evidently absurd about them. The Newtonian universe of the eighteenth century was filled with invisible fluids that were the particular favorites of one or another Enlightenment philosopher: animal heat, organic molecules, inner mold, ether, phlogiston, and more. Who was to say whether Mesmer's particular fluid was more or less plausible than any of these?[21]

In fact, it could be that Mesmer's animal magnetism had the edge on some of these other invisible forces, because it was not merely an abstract theory dreamt up by an Enlightenment philosopher; rather, it was a physical experience. Patients undergoing treatment reported feeling powerful electriclike sensations shooting through their bodies, fluid moving up and down their limbs, intense tingling, heat, and cold.[22] Indeed, the physical sensations happened sometimes in apparent defiance of a subject's will. In the polite society of mid-nineteenth-century England, Thomas Carlyle's wife, Jane, told the following story of her encounter with a lower-class mesmerist in a letter to her uncle:

> I looked him defiantly in the face as if to say, you must learn to sound your H's Sir before you can produce any effect on a woman like me! and whilst this or some similar thought was passing thro' my head—

> flash—there went over me from head to foot something precisely like what I once experienced from taking hold of a galvanic ball—only not nearly so violent—I had presence of mind to keep looking him in the face as if I had felt nothing and presently he flung away my hand.[23]

If a strong-minded upper-class lady could not stop herself from physically feeling the magnetic energy directed at her by an ill-speaking working-class man, could anyone really doubt its reality?

In fact, the answer to this question was yes; for at the time, not everyone was persuaded that feeling was believing. Mesmer himself felt the brunt of this kind of skepticism in the late 1770s when he was accused of fraud—and maybe something worse—in his treatment of one particular patient. Maria Theresa Paradis was a young pianist from a prominent family who had been blind for fifteen years, even though doctors could find no organic cause for her condition. Mesmer went to work on her and claimed to be achieving some real success. She began to say she could see, but then she insisted she could see only when Mesmer was in the room and near her physically. What kind of cure was that, then? As suspicion grew, the girl's father—court counselor to Empress Maria Theresa, after whom his daughter had been named—weighed in with a different concern. There was something improper about the relationship between Mesmer and the young woman; she was inappropriately attached to her doctor and had now turned against her family. In the end, the court counselor forcibly removed his daughter from Mesmer's care, but by this time the case had become notorious and Mesmer's reputation was seriously endangered.[24]

Mesmer concluded that it was time to leave Vienna. He chose Paris, known for its receptivity to novel ideas, and settled there in 1778. Initially, this looked like an inspired decision, as he quickly attracted the attention of a new set of fashionable followers, including Queen Marie Antoinette. Some of the new recruits were determinedly earnest. The Comte de Montlosier, for instance, was converted early and continued over the years to declare his belief that Mesmer's system would "change the face of the world. . . . No event, not even the Revolution, has provided me with such vivid insight as mesmerism."[25] Others, though, were clearly attracted more to the novelty and spectacle of it all. The mid-nineteenth-century author Charles Mackay, in his thoroughly opinionated and witty history *Extraordinary Popular Delusions and the Madness of Crowds*, painted the following picture of the scene around Mesmer in those years:

> *"Was ever any thing so delightful!"* cried all the Mrs. Witterley's of Paris, as they thronged to his house in search of pleasant excitement; *"So wonderful!"* said the pseudo-philosophers, who would believe any thing if it were the fashion; *"So amusing!"* said the worn-out debauchés, who had drained the cup of sensuality to its dregs, and who longed to see lovely women in convulsions, with the hope that they might gain some new emotions from the sight.[26]

If the scene portrayed by Mackay even approximated the actual mood of intrigue surrounding Mesmer's work in Paris, then it is hardly surprising that some people, in the medical profession and elsewhere, should have begun to clamor for him to be investigated. In the spring of 1784, King Louis XVI authorized two royal commissions made up respectively of members of the Royal Society of Medicine and of the Royal Academy of Sciences to carry out the job.[27] The work of the first, strictly medical, commission was largely forgotten soon after its report.[28] However, the work of the second commission was to prove much more influential. The membership of this Commission reads like a *Who's Who* of natural philosophy at the time, including for example, Benjamin Franklin (at the time serving as American ambassador to France), the astronomer Jean Bailly (who had computed the orbit of Haley's Comet and studied the moons of Jupiter), the chemist Antoine Lavoisier (who discovered oxygen), and the physician Joseph-Ignace Guillotin (whose main claim to fame, being embodied in his name, hardly requires further mention).

These men oversaw a series of rather clever trials that together served to demonstrate to their complete satisfaction that there was no evidence that an animal magnetic fluid was responsible either for the convulsive crises or the resulting cures of Mesmer's patients.[29] In one trial, for example, a particular tree in a garden was magnetized, and a blindfolded boy was led up to four trees in succession. At the first tree, he began to sweat. By the time he had been brought to the fourth tree, he had fallen into a state of violent convulsions—but all without having come anywhere near the tree that actually had been magnetized! What was responsible for his behavior? Evidently not the presence or absence of any invisible fluid, the investigators argued, but merely the power of his imagination: "The imagination without the aid of Magnetism can produce convulsions," and much more besides; but "Magnetism without the imagination can produce nothing." The commission concluded that the magnetic fluid did not exist. Mesmer was wrong.[30]

These conclusions are often taken to represent an early triumph of the scientific method over gullibility, but this fails to do justice to the situation. We are not dealing here simply with a desire to expose fraud, as in the case of Marthe Brossier; nor are we dealing with skepticism toward a particular explanatory framework that could be replaced by another, better one, as in the case of Gassner. Instead, the second commission's conclusions inaugurate a style of skepticism we have not yet seen: skepticism toward the psychological.

What I mean by this, to begin, is that the commissioners' skepticism was not directed toward the phenomena themselves. The commissioners freely conceded that the treatments they had observed had the capacity to produce powerful bodily effects in some people—convulsions, tremors, and more; and they were even open to the possibility that some of these effects might be of a therapeutic nature. But they found that the cause of these effects lay not in the physical but in the mental realm; not in Mesmer's supposed magnetic "fluids" but rather in a faculty of mind they called the "imagination." In stating this conclusion, however, they were not proposing to replace Mesmer's schema with an alternative one (as Mesmer had previously done with Gassner). What they were doing was dismissing those effects as unworthy of explanation altogether. Imagination, these commissioners knew (following a considerable tradition in the eighteenth century), was the enemy of rational enquiry—a quixotic, irrational, and poorly controlled faculty of the mind. Its fancies, especially ones so powerful they could spread across the population, were a danger to clear thinking because they were not grounded in truth. Scientific methods were thus required, not to understand them, but to unmask them for the unruly, dishonest things they really were.[31]

Later, d'Eslon, the mesmerist who was the actual target of the commis-

Anonymous French cartoon, 1795, one of a number of satirical and salacious images that appeared in the French press in the wake of the 1784 report repudiating the authority of Mesmer's treatment. The mesmerist is shown here with the ears of an ass, a stock symbol of folly, brandishing a "magic finger," while his female patient appears to be having visions of an angel chasing a siren. Note also her hand resting suggestively on her groin. Behind the mesmerist are piles of coins, indicating a lucrative practice. The caption reads "An Ape is always an ape," suggesting that the image refers to Mesmer's foolish imitators rather than to Mesmer himself.

sioners' investigation (see note 27), complained that the commissioners made no effort to define the "imagination" to which they attributed the magnetic effects. He had missed the point. For the commissioners, the mere fact that a treatment worked—at least on some level, and some of the time—was not sufficient grounds itself to take it seriously. In the words of French philosopher Isabelle Stengers, "the suffering body [alone] is not a reliable witness" to the validity of a treatment. "It can happen that it will be cured for the 'wrong reasons.'"[32] This conclusion helped set the terms for the emergence, a century later, of a new narrative about suggestion whose cures simultaneously acted to unmask the suffering patient as a victim of self-deception.

From convulsions to trance

Before that could happen, however, other things had to happen first! Following the report of the second commission, mesmerism did not disappear, but it did begin, more and more, to dissolve its early alliances with representatives of academy science and medicine.[33] Instead, its practitioners and advocates defiantly took their case directly to the people. These were the years that saw the rise of itinerant stage magnetizers, men who traveled from city to city (often accompanied by trained subjects) giving performances that blurred distinctions between public demonstration and public entertainment. Even as this was going on, however, the experience of mesmerism itself—what people felt and did under the alleged influence of animal magnetism—began to change.

The stage magnetizers themselves were not directly responsible for this change, though they would do more than anyone else in spreading the new formula across Europe. The immediate cause seems to lie instead in a growing sense of unease with the so-called magnetic convulsions of mesmerism. Prior to Mesmer, the ability to provoke convulsions on command had been one of the most common ways an exorcist flushed out a demon and demonstrated his control over it. And although Mesmer, in his initial secularization of the possession ritual, denied convulsions their theological meaning, he never denied their importance. For him, they remained the climax of the healing drama; in this he was in accord with the exorcists.

Times, though, were changing, and the notion that it was necessary to succumb to violent convulsions as a prelude to healing was increasingly out of step with emerging new sensibilities. Some patients of Mesmer had already expressed their fear and dislike of the convulsions; and the second commission, charged with investigating the scientific merits of Mesmer's theory, devoted a significant portion of its text to expressing its authors' troubled feelings about this aspect of the treatment. Some of the convulsions lasted for more than three hours, they exclaimed, during which time patients vomited violently, shook all over, had trouble breathing, and shrieked, sobbed, and laughed in turns. "Nothing is more astonishing than the spectacle of these convulsions," they concluded. "One who has not seen them can have no idea of them." They warned against the "deplorable consequences" of such spectacles, and even against the witnessing of them (since witnessing seemed to provoke imitation in other patients).[34]

That was in the public report. In a "secret report" intended for the king's eyes only (it was first published in 1800), the commissioners offered a different—moral and social—reason to worry about the convulsions. At least some of the antics they had witnessed suggested to them that some of the convulsions were actually sexual in nature—that women were having orgasms in the treatment room.[35] It had long been well appreciated that the magnetic treatment involved quite close contact between the (male) magnetizer and the (often female) patient (stroking the thighs, trailing fingers just over the face, breathing on the eyes, etc.). Now, though, the commissioners were suggesting real reasons for concern about the effects of these mesmeric intimacies on public manners and morals.

All this anxious talk about convulsions and their multiple dangers came at a time of growing anxiety (not unwarranted) among the upper classes of European society about stresses, strains, and—possibly—convulsions in the body politic. Increasingly, anything that might inflame popular passions came under scrutiny. These were years that saw, among other things, the first efforts to outlaw or restrict blood sports like bear baiting and to abolish the public executions that were increasingly felt to increase rather than reduce crime (they were abolished in France in 1780 and in England in 1783).[36] Taken together, these general trends provide a framework for us to make sense of why the mesmeric crisis or healing convulsion gradually gave way to a new mesmeric narrative organized around a new climactic centerpiece: the descent into a trance state that would be called "magnetic somnambulism."

The figure with whom this shift is generally identified was a gentle aristocrat named Armand-Marie-Jacques de Chastenet, the Marquis de Puységur (usually referred to in the literature as simply "Puységur"). He had made a visit to Mesmer's salon in the 1780s and was impressed with the therapeutic results. A bit like the commissioners, however, he had been dismayed by the violent spectacle of the convulsions Mesmer deemed necessary for a good outcome. When he returned to his estate of Busancy, near Soissons, Puységur began mesmerizing several more or less willing servants (most famously, a young peasant named Victor Race). But he resisted generating convulsions in his subjects. He had decided they were not essential to the magnetic crisis and blamed their frequency in the past on the fact that Mesmer and his followers had tended to work with groups of patients too large to look after properly. When treatment was undertaken one-on-one and in a calming environment, he insisted, the crisis unfolded in a very different way.[37]

Specifically, subjects sank into a strange state of mind somewhat akin to sleepwalking; Puységur called this magnetic somnambulism. While in this state, they spoke very frankly about their innermost thoughts, seemed to have remarkable control over their bodily processes, and responded to instructions from the magnetizer without hesitation. Not only were subjects typically more intelligent in the somnambulic state than in their normal waking state, but some of them also seemed to develop paranormal knowledge and abilities. It would become common for Puységur's subjects to diagnose and predict the course of their own disorders and the disorders of others. Some were also apparently able to read the mind of the magnetizer and respond to his commands "at a distance." This latter ability was linked to the creation of an energetic connection between the magnetizer and the patient, during which mind melded with mind and will submitted to will. Puységur called this phenomenon "intimate *rapport*." He compared it to the attraction seen in mineral magnetism between an iron rod and a magnetic needle: the somnambulist obeys the mesmerist just as the needle obeys the rod.

When they were returned to their normal state, subjects seemed to remember nothing of what had just transpired (in contrast, in the magnetic state they remembered everything from their normal state).[38] How did Puységur make sense of all this? While he didn't actually deny that the fluid had a hand in producing this new sort of experience, he also insisted that the *will* of the magnetizer was the key piece of the puzzle: "Animal magnetism does not consist in the action of one body upon another," he wrote in 1807, "but in the action of the thought upon the vital principle of the body."[39] A psychological principle, fashioned partially metaphorically on the physical idea of magnetic action and attraction, had entered the discourse properly for the first time.

Within a few years of Puységur's modeling this style of mesmerism, something remarkable had happened: virtually all patients subjected to magnetic manipulations across Europe had ceased to convulse and instead had begun falling into somnambulic states. On the brink of the French Revolution, mesmerism had survived by transforming itself from an experience marked by violent convulsions en masse to an experience marked by quiet one-on-one obedience to an authority figure.

This is not to say that the specific structure of this new-style mesmerism was nothing more than a political adaptation to changing times. On the contrary, no less than the convulsions it replaced, magnetic somnambulism—and the special abilities people supposedly acquired when in such a state—gives every indication of being a secular adaptation of a part of the Catholic exorcism ritual that had been ignored by the first generation of mesmerists, specifically, that moment in the ritual when the exorcist commanded the demon to speak and reveal its supernatural powers and secret knowledge.[40] We can partly understand why Mesmer and his immediate followers would have ignored this facet of the possession narrative: the Newtonian theory of fluids and forces on which mesmerism was based offered no real resources for interpreting it satisfactorily.

Just a few decades later, however, as magnetic somnambulism spread, the intellectual scene looked rather different. New natural philosophies that scholars would later conventionally designate as "romantic" had emerged, arguing that the mind was not the lawful, predictable, and transparent entity assumed by the mechanistic philosophers. It was creative, unpredictable, and mysterious—and this was all right! In fact, it was something worthy of celebration.[41] Even the imagination so distrusted by eighteenth-century rationalists was widely rehabilitated. In this context, magnetic somnambulism could also come into focus as one more product of the mysterious workings of the mind and the self. The irrational layers of the mind, it began to seem to some, just might be worth studying and taking seriously after all.[42]

These new ways of thinking about the mind were not of mere academic or philosophical interest. They had, at least in the eyes of some, immediate practical implications. In 1844, the *New York Sun* published a story about two women friends who, while their husbands were both in the city to purchase goods, proceeded to mesmerize each other for their own amusement—and unexpectedly gained clairvoyant powers. One "saw" her husband gambling away the family funds; the other "saw" hers buying expensive gifts for a secret sweetheart. Confronting the startled men with this knowledge on their return home, the wives both decided to forgive and forget, but the author of the article intoned the moral: "Let husbands from home be cautious how they act, and keep in the right path, for by this new mesmeric discovery, wives can keep a watchful eye on all their movements, and nothing can be concealed from them in applying to this great moral discovery for the means of knowledge and detection."[43]

The doctors get involved

In the years after Puységur, nurtured in the hothouse atmosphere of romanticism, a protopsychology of magnetic somnambulism grew, rich with mystical implications. Words like "magnetism" and "rapport" (which had originally referred to responses by people to a supposedly real fluid) began to take on the interpersonal meanings of attractiveness and emotional connection with which we associate them today. By the mid-nineteenth century, the fluid had become almost incidental to the story and drama of magnetizing. In this sense, the invention by the Manchester physician James Braid of a new psychological procedure for invoking waking sleep—"hypnotism" (originally "neurypnotism")—was more an evolutionary than a revolutionary moment.[44]

Braid's interest in hypnotism (as we must now call it) had been sparked by close observation of several public demonstrations carried out by a visiting Swiss mesmerist named Charles Lafontaine. To his own surprise (he had gone to these demonstrations as a distinct skeptic), Braid found himself forced to conclude that Lafontaine's manipulations were producing "real phenomena." The mesmerist was wrong, however, to attribute those phenomena to animal magnetism.[45] Experiments carried out by Braid suggested they were instead produced through "fixation of the mind and eye." That is, they could be brought on by making subjects concentrate on a single monotonous idea while staring at some designated object. Mesmerists like Lafontaine had long unwittingly employed this method through their habit of making monotonous passes up and down the bodies of their subjects. Because they did not understand what they were really doing, however, their results had been inconsistent, and the medical profession had failed to take any of their claims seriously. Braid believed his discoveries provided an entirely new basis for serious investigation:

> I have now entirely separated Hypnotism from Animal Magnetism. I consider it to be merely a simple, speedy, and certain mode of throwing the nervous system into a new condition, which may be rendered eminently available in the cure of certain disorders. I trust, therefore, it may be investigated quite independently of any bias, either for or against the subject, as connected with mesmerism.[46]

Over the next several decades, a thin stream of medical researchers would indeed be tempted to take a fresh look at hypnotism. Most of these followed Braid in seeking to distinguish this phenomenon from its discredited relation, mesmerism. Nevertheless, the larger medical and scientific community generally failed to recognize such niceties of difference and tended instead to dismiss the entire drama of provoked waking sleep—however achieved—as chicanery and unworthy of scientific and scholarly attention.

Then, the French neurologist Jean-Martin Charcot, one of the most charismatic and internationally respected figures in the medical profession at the time, proposed that it was time to take hypnosis seriously. In a lecture before the French Academy of Sciences in 1882, he described hypnosis as an artificially induced modification of the nervous system whose medical interest lay not—as Braid and others had thought—in its potential therapeutic applications, but in the fact that it could *only* be produced in patients suffering from hysteria. It consisted of discrete phases—catalepsy, lethargy, and somnambulism—each of which could be identified by special physiological signs and provoked by stimulating the nervous system in specific, differentiable ways.[47] Catalepsy was said to be characterized above all by loss of autonomous will and muscular rigidity; subjects could be put into a range of positions and poses, and would hold each indefinitely (this capacity gave rise to the parlor trick still sometimes practiced today in which a hypnotist stands or kneels on the rigid body of a hypnotized person suspended between two chairs). Lethargy was described as a state of flaccid unconsciousness and heightened muscular reactivity. Somnambulism was comparable to the state we think of today as the hypnotic trance, in which subjects could see, speak, and move but could only initiate actions demanded by the hypnotist and did so without apparent resistance.

Jean-Martin Charcot (1825–1893), photograph by Nadar, 1890. *© Bettmann / Corbis*

Described in this bare-bones way, Charcot's 1882 lecture might not appear to be radical stuff, but it changed everything. Across France and (to a lesser extent) the rest of Europe, researchers felt they had finally been given permission to admit their interest in this long-shunned phenomenon. A flood of new investigations appeared. By transforming hypnosis into an induced (hysterical) pathology that followed regular, physiological laws, Charcot had succeeded in doing two things: giving an aura of respectability to the subject; and staking a clear claim to the medical profession's exclusive competency to deal with it. The effect, as French psychologist Pierre Janet would later recall, was stunning; a true tour de force of persuasion.[48]

Charcot's link between hypnotism and hysteria is the key to understanding his rhetorical success. Hysteria, a disorder characterized by a bewildering and baroque array of symptoms (convulsions, paralyses, tunnel vision, color blindness, patches of anesthesia, incessant coughing, tics, feelings of choking), had long been the bane and frustration of clinicians across Europe. In the 1870s, following a remarkably successful decade of work elucidating a range of other brain disorders, Charcot had decided it was time to make hysteria lie down and behave. Using a well-tested approach that involved correlating symptoms with anatomical evidence of disease, he resolved to bring it into focus for himself and his medical colleagues.

The strategy worked—or at least it half worked. Brain dissections in the laboratory produced no clear results. However, underlying patterns for the permanent symptoms, the so-called stigmata, of hysteria began to emerge. The convulsive crises associated with the disorder also turned out to be far less chaotic than they had previously seemed, unfolding in four stages that, as Charcot put it, followed one another with "the regularity of a mechanism."[49] To get an accurate visual record, assistants were set to work sketching pictures of designated patients. Not long after, a photographic studio was installed in the hospital (a major innovation at the time), and photographic records began to be made instead. Colleagues said that the camera was as crucial to the study of hysteria as the microscope had been to the study of histology. The camera, they said, "did not lie," and it therefore provided the evidence Charcot needed to prove, as he put it, that his laws of hysteria were "valid for all countries, all times, all races," and were "consequently universal."[50]

In a way, it was a short step from the camera to hypnosis for Charcot, because both tools served the same investigatory ends.[51] More precisely, they worked in synergy. Hypnosis allowed researchers to manipulate the nervous system of hysterics in a controlled fashion; the camera allowed them to capture the resulting data for later study. Photographs of "hypnotizable hysterics" in various poses and states began to fill the medical journals, and Charcot himself became famous for his public demonstrations of these states and poses, executed with great showmanship before audiences of students, colleagues, and select members of the general public.

Later, critics would rather cruelly—but with some justice—compare Charcot's public lectures to the stage performances of the lay magnetizers Charcot and his colleagues had taken to condemning as dangerous amateurs. Nevertheless, Charcot's understanding of hypnosis was something altogether different. Concepts like rapport played no part in his thinking. As he saw it, his exhibitions were not interpersonal dramas, but demonstrations of a tool capable of revealing certain laws of physiology under pathological conditions. When he performed, he was always in the role of the scientist who stood outside the system he was studying, pushing buttons and pulling levers simply (or so he thought) in order that the facts of nature might be revealed.

Putting together the pieces: the invention of suggestion

For several years, this strategy seemed to yield brilliant results. Then, in the mid-1880s, a rival of Charcot, Hippolyte Bernheim, doctor of internal medicine at the University of Nancy, appeared on the scene and began to make trouble. In particular, through hypnosis demonstrations of his own, Bernheim began to present evidence that interpersonal dramas might have had the upper hand in Charcot's research program after all.

Bernheim had become interested in hypnosis through the work of a modest country doctor named Ambroise-Auguste Liébeault, who since the 1860s had been quietly curing peasants of various disorders using a form of hypnosis partly adapted from the ideas of James Braid. After briefly requiring his patients to fix their eyes on his own (a variation on Braid's method), Liébeault would tell them in a gentle but emphatic

Hippolyte Bernheim (1840–1919). Source: *Nos Maîtres.*

voice that they were now to sleep (sometimes physically pushing down their eyelids if they were slow to comply). This generally succeeded in producing a somnolent state. Liébeault would then "affirm" to them in no uncertain terms that they were feeling better, that their symptoms were being relieved, and so on. When they awoke, many of them found that it was so.[52]

Originally skeptical, Bernheim became a convert to the old doctor's homey methods and spent years at the University of Nancy refining the approach and developing a theory of the mechanisms behind it. The results of that research were published in 1884 in a book that, as historian Alan Gauld has noted, "was ultimately to have a greater influence than anything which had appeared in the hundred years since Puységur's *Mémoires*."[53] The book was titled *De la suggestion dans l'état hypnotique et dans l'état de veille* (*Suggestion in the hypnotic state and in the state of waking*).[54]

"Suggestion." We finally encounter this word. But what did Bernheim mean by it? His formal definition described it, rather tautologically, as the "influence provoked by an idea suggested and accepted by the brain." By this he meant a process in which an idea is accepted in such a way as to lead to "ideomotor and ideosensory automatisms"—i.e., patients feel and behave in a way consistent with an implanted idea without reflecting on its sense or plausibility.

Bernheim believed a tendency toward "suggestibility" was natural to human beings. Nevertheless, it often worked better if a patient could be lulled into a state of reduced vigilance. That state itself could be attained through suggestion; and once attained, its only unique characteristic was that it acted to further enhance suggestibility. "The idea makes the hypnosis," Bernheim insisted; "it is a psychical and not a physical or a fluid influence which brings about this condition."[55]

This understanding, of course, could hardly have been more different from that of Charcot, and it provided Bernheim with a perspective from which to thoroughly deconstruct Charcot's approach. Using suggestion, Bernheim showed that he could reproduce all the symptoms and stages that Charcot had identified as the objective signs of hysteria and hypnosis—and then proceed to change them or make them disappear. All the dramatic neurological effects produced by Charcot, he said, had

Pierre André Brouillet, *Une leçon clinique à la Salpêtrière*, 1887. Charcot is seen here demonstrating that the symptoms of hysteria are as real as those of any organic disease. The patient, Blanche Whittman, is about to experience an attack typical of "major hysteria." Facing her is a painting that shows a patient undergoing the same kind of attack, in this way virtually modeling the behavior expected. To later critics, the presence of this painting within a painting has become a kind of unwitting symbol of the way in which Charcot's entire approach to hysteria was actually grounded, not in physiology but in the power of suggestion. *© Leonard de Selva / Corbis*

nothing to say about how the nervous system really worked. They were simply ideomotor and ideosensory responses patients made to cues—suggestions—unwittingly given to them by the doctors. Charcot and his colleagues, so scornful of the chicanery they associated with the showmen magnetizers of their time, had in the end managed to do nothing more than create an elaborate theater of illusions themselves.

This critique devastated Charcot's program. Not since the concept of "imagination" was used in the late eighteenth century to demolish the credibility of Mesmer's animal magnetism had a psychological construct proved so effective in undermining the credibility of physicalist thinking in medicine. For a while, Charcot and his students tried to fight back against the accusations leveled against them, through an appeal to common sense and the solidity of the master's methods:

> It has been said that it [so-called major hysteria] exists only at the Salpêtrière, as if I have created this condition by my own willpower. What a marvel this would be if I could, in fact, fabricate illnesses according to my whims and fantasies. But in fact I am a photographer. I describe what I see.[56]

In the end, however, these and similar protestations were of little use. Charcot and his students were increasingly ridiculed, judged as scientists who had been done in by their own psychological naïveté. When Charcot died in 1893, his reputation, at least in this area, was in tatters, and his students, looking to rescue their careers, scattered. In 1901, one of Charcot's previously most loyal students, Joseph Babinski, read a paper on hysteria before the Neurological Society of Paris that represented a kind of apotheosis of the skepticism toward the psychological that Bernheim's "suggestion" had spawned. The paper was a mea culpa on behalf of both himself and the whole approach of the school he had once defended. They had been wrong, he said; hysteria had nothing to do with brain physiology and anatomy. It was instead a will-o'-the-wisp of the mind alone, nothing more than the sum total of symptoms that could be created by suggestion and removed by suggestion. And in an echo of the conclusions reached by the commission that had investigated Mesmer more than a century earlier, Babinski concluded that medicine need pay no more attention to hysteria.[57]

An irony in all these reactions is that Bernheim himself was not skeptical of the psychological in the way that the commissioners who had used imagination to debunk Mesmer's animal magnetism had been. He did not believe that suggestion was only good for producing deceptive forms of pseudophysiology. On the contrary, when he was not busy attacking the Charcot school, he was happy to emphasize the positive therapeutic potential of suggestion, and not only for disorders that were hysterical in origin. Nevertheless, Bernheim's call for medicine to develop suggestion as a therapeutic method would always be far less successful than his debunking work.

The reasons for this extend beyond any inbred skepticism toward the psychological felt by Bernheim's medical peers. Looking behind the veneer of his technical language, it is clear that Bernheim was advocating the therapeutic use of a process that looked at once irrational and unethical, a process that brought to mind all of the old mesmeric stories in which the mind's powers to influence the body were bought at the price of the patient's free will. The public had never really stopped believing in those older stories—not least because they were reinforced by tantalizing and tawdry fictional works such as the late-nineteenth-century blockbuster by Englishman George du Maurier, *Trilby*. The novel tells the story of an evil-minded hypnotist who turns a beautiful, unsuspecting young girl into a singing automaton and his sex slave. First published serially in *Harper's Monthly* in 1894, it gave the English-speaking world a new word: "Svengali."[58]

Trilby, of course, was fiction; but the new ideas about suggestion seemed to lead to the conclusion that it could have been fact. "The hypnotists of the Nancy school rediscovered and gave general currency to the doctrine that the most essential feature of the hypnotic state is the unquestioning obedience and docility with which the hypnotized subject accepts, believes, and acts in accordance with every command or proposition of the hypnotizer," declared the anonymous author of the 1911 entry on "suggestion" for the *Encyclopedia Britannica*.[59] The German physician Moriz Benedikt warned against the use of hypnotic suggestion in a medical context—it could turn a patient, he warned, into a "pliable tool" of his or her physician.[60] His French colleague Jules Liégeois went even further: in the hands of the unscrupulous, suggestive techniques could be bent to criminal ends; they might even be used to "subvert the army."[61]

Ironically enough, in the end it was the armies of Europe and the United States that found the most use for suggestive techniques as they attempted, during two world wars, to quickly put the pieces back together for the many soldiers who had broken down in battle and subsequently been diagnosed as suffering from shell shock or battle fatigue. If classical hypnotic suggestion was generally regarded as being too coercive for normal civilian medical practice in the early twentieth century, its no-nonsense insistence on submission was part of what made it attractive for use in a military setting. In the 1946 John Huston documentary of war trauma *Let There Be Light,* army psychiatrist Benjamin Simon employs a highly authoritarian form of (chemically facilitated) hypnotic suggestion to treat the soldiers under his care. Like a charismatic exorcist, he commands a paralyzed soldier to walk (and we watch the young man rise and do so) and a badly stuttering second soldier to talk (and we watch as the young man, weeping with joy, realizes he can speak again). When Simon has to cure the amnesia of a soldier who does not even remember his own name, the process is tough—the soldier is commanded under hypnosis to go back in his mind to where the troubles began and becomes visibly distressed. But Simon stays in control and makes clear that he will brook no disobedience: "You're going to remember it all. You're going to remember about Okinawa. You're going to remember about the shells and the bombs, but they're gone." The film's narrator then soothingly explains: "Under the guidance of the psychiatrist, he [the soldier] is able to regard his experience in its true perspective as a thing of the past which no longer threatens his security. Now he can remember."[62]

Placebos and the power of suggestion

If suggestion was openly deemed a good treatment for traumatized soldiers used to taking orders, its usefulness was also recognized—though much less openly—for ordinary civilian patients also long accustomed to doing whatever the doctor told them. Given the concern at the time about the ethical propriety of hypnosis, however, doctors working in a civilian context preferred to work their suggestive effects by indirect means—using fake pills and potions—and referred to the results by a new name: the placebo effect.

"Placebo" itself is a very old Latin word meaning "I shall please." In fact, it is used to open the Catholic Vespers for the Dead. In medieval times, the Vespers for the Dead were frequently sung by hired mourners who were often seen as sycophants who wept crocodile tears on behalf of a family. This gave "placebo" the rather odious meaning of flatterer or toady. Transferred as a concept to medicine sometime in the early nineteenth century, placebos became mollifying objects that doctors gave to patients for whom they could otherwise do nothing or who were suspected of having nothing really wrong with them. Bread pills, tonics, powders, tinctures—all these things were used by doctors, though always with some sense of defensiveness and ambivalence.

In 1903, Richard Cabot, Harvard Medical School professor and a doyen of American medicine, confessed: "I was brought up, as I suppose every physician is, to use *placebos*, bread pills, water subcutaneously, and other devices for acting upon a patient's symptoms through his mind."[63] He was, however, willing to make such a confession only because he had now decided that such practices were wholly unacceptable:

> Every placebo is a lie, and in the long run the lie is found out. We give a placebo with one meaning; the patient receives it with quite another. We mean him to suppose that the drug acts directly on his body, not through his mind. . . . If the patient finds out what we are doing, he laughs at it or is rightly angry with us. I have seen both the laughter and the anger—at our expense. Placebo giving is quackery.[64]

It might have been quackery in the eyes of Cabot, but most physicians still felt it was a form of quackery too useful to give up. Moreover, the late-nineteenth-century narrative about "the power of suggestion" provided a new rationale for continuing to practice this "humble humbug."[65] What if it were possible to think about the placebo less as a form of quackery and more as a form of suggestive psychotherapy? This was the defense of the practice offered in 1945 by Oliver Hazard Perry Pepper, a prominent Philadelphia-based physician in the interwar period.[66] Pepper noted that so far as he knew, his was the first article on placebo use ever to appear in the published medical literature. Sometimes, he then went on, a doctor had nothing to offer except the force of his authority and the comforting rituals of his trade. Why should he withhold from patients any consolation such things might be able to offer, especially in those cases of terminal illness where conventional medicine had exhausted its options? "The human mind is still open to suggestion," Pepper noted, "even in these modern and disillusioned days. The sympathetic physician will want to use every help for these pathetic patients and if the placebo can help, he will not neglect it. It cannot harm and may comfort and avoid the too quick extinction of opiate efficacy."[67]

Such sentiments, however, were by 1945 already the talk of a fading generation. Medicine was in the process of changing, transforming itself into a practice rooted in the laboratory. It was also in the first stages of claiming for itself an entirely new arsenal of pharmaceutical interventions, from new analgesics to new antibiotics. These larger social, economic, and institutional changes produced conditions that would turn the vision of placebos as little packets powered by suggestion into something that was less a "humble humbug" and more a major threat to the progress being made on all of these fronts.

A 1955 paper by Henry Beecher, "The Powerful Placebo," catalyzed and embodied this sea change in understanding. The paper had two goals: on the one hand, it argued for a new understanding of placebos and their effects; on the other, it urged the adoption of a new research methodology to remove (among other things) the distorting influence of placebo effects from clinical research settings.

For Beecher, placebos were not harmless humbugs, psychotherapy by other means offered to patients who could not otherwise be helped. Analyzing the combined results of fifteen clinical trials that used inert tablets as placebo controls, Beecher determined that the effect of these placebos was profound: on average, placebos affected about a third of patients in the placebo group (though there was great variability across the trials), and in certain instances produced "gross physical change[s]" that could "exceed those attributable to potent pharmacological action."[68]

Recognizing that placebo effects were powerful, however, was not the same thing as saying they were therapeutically useful. On the contrary, Beecher was clear that the effects seen had all the spurious qualities long associated with bodily phenomena produced by suggestion. In the modern era, medicine could not afford to risk being misled. Here, Beecher came to the second goal of his paper, one that was effectively political. Medical research needed to protect itself against the powerful placebo by adopting a cutting-edge methodology that would come to be called the randomized controlled trial (RCT). This method involved randomizing the patients for a study into two groups, in considerable part so that each group would have roughly equal numbers of "placebo responders." One of the groups was then given a sham version of a new medication (without knowing it was sham); the other received the real treatment. Because of randomization, it was supposed that placebo effects—suggestive effects—would operate in both groups at approximately the same level. Any difference in responses to the medication seen between the two groups was supposed to reveal how effective the medicine in question really was.[69]

And this was, for some time, where matters rested. The mind's capacity to create pseudosymptoms—recognized as a spoiler of clear observation in medicine since the time of Mesmer—would now be contained and neutralized through the ritual of the clinical trial. It would be another three or four decades before some people in American medicine would begin wondering whether the sugar pills used in clinical trials might be producing effects that were more real—i.e., more biologically and therapeutically significant—than may at first have been apparent. As questions of this sort began to be asked, the placebo effect ceased to be a drama about "the power of suggestion" and became instead a drama about "the power of positive thinking" (chapter three).

Today, the existence of a cluster of upbeat narratives about the mind's capacity to affect the body's health might lead to the impression that the skeptical vision of mind-body interactions embodied in "the power of suggestion" has been marginalized, if not made historically obsolete. That impression would be wrong. The specter of the exorcist-Mesmerist-Svengali manipulating and deceiving minds and bodies through suggestion retains a powerful hold on our contemporary imagination, as does the specter of our mind playing tricks on us by making us have experiences that are not real. Those most likely to invoke these specters are people who otherwise tend not to hold much truck with so-called holistic, alternative, and mind-body therapies. Many might even say they do not give much credence to the mind-body connection; they don't believe that all those positive-attitude mantras, meditation practices, support groups, and psychotherapies designed to uncover unconscious traumas actually do patients any good. They do, however, believe in suggestion, because the essence of suggestion is that it is not a mind-body therapy so much as it is the charlatan's counterfeit currency. The following quote, from an article published in 1998 in a journal of medical ethics, makes the point plainly:

> Dr. Miracle misperceives his role as a physician by viewing himself as the sole source of healing. His actions do not engage the patient, either diagnostically or therapeutically, as a partner in the effort to restore Mr. Misery's ability to live his life as he chooses. Dr. Miracle apparently regards his role as that of a conjurer or shaman whose tools and methods require mystery for their effect. Note that Dr. Miracle does not use patience, compassion, trust, or any other of our usual human means of interacting with patients: Instead, Dr. Miracle, befitting his name, chooses magic.
>
> "And what is so wrong with magic?" Dr. Miracle replies. "It got the job done." What is so wrong, we counter, is that this form of magic is an illusion performed more for the benefit of the magician than for the patient; the metaphor at work here is that of the exorcist trying to trick the pain into leaving the body. This is a thoroughly inadequate model of medicine for our age."[70]

For this author, who is fundamentally skeptical of all things psychological, the only legitimate alternative to a Dr. Miracle is a biomedicine based in solid physiology. But is that the last word on the matter? Since the late nineteenth century, there have been some who have insisted that the answer is no; that one can take the mind's effects on health seriously and even harness them in the service of healing without resorting to either deception or coercion. In this sense, they have said, talk about "the power of suggestion" was not enough. Mind-body medicine was in need of some fresh narratives, narratives that emphasized the healing power of truth telling over deception, and the potential for authentic, lasting cures over temporary, illusory ones. Such narratives would not be long in appearing.

Chapter Two

THE BODY THAT SPEAKS

You know I can't express emotions. I internalize, I grow a tumor.
—*Woody Allen,* Manhattan

In the early 1980s, a middle-aged woman named Gil (a pseudonym) sat down for an interview with a sociological researcher. For some years Gil had suffered from debilitating rheumatoid arthritis, and now she was being invited to explain the sense she made of her experience. Why did she think this had happened to her, she was asked. She reflected. Maybe, she said, it was just part of the fact that we were all fallen children of God. "It's the old Adam, we've all got to be ill." No, she then corrected herself; perhaps it was just a simple mechanical thing: "You're bound to get worn out parts, like cars." Having offered in turn a theological and a quasi-mechanistic explanation for her illness, she reflected a third time, then leaned forward and confessed to the researcher a deeper suspicion about the cause of her suffering:

> Mind you, I sometimes wonder whether arthritis is self-inflicted . . . not consciously. You know, your own body says, "right, shut-up, sit down, and do nothing." I feel very strongly about myself that this happened to me, that one part of my head said, "if you won't put the brakes on, I will."[1]

What kind of logic is operating here? What could this woman have meant by saying that her disease was "self-inflicted," but "not consciously"? What kind of body is it that can use the symptoms of arthritis to tell a person to "shut-up, sit down, and do nothing," literally forcing her to do as it says because her arthritic joints won't allow her to move fast anymore?

The answer is: a talking body. A body that can do what Gil thinks her body has done to her is one that can converse with its owner, conveying messages to her about things going on in her life that she is having difficulty consciously confronting. This strange body stands at the center of a narrative about the dangers of repression and the healing power of confession that I call "the body that speaks." It is a narrative that is historically indebted, more than anything else, to the work of Sigmund Freud. People like Gil who invoke this idea, however, generally do not think about or perhaps even know that. For them, the idea that bodies might be able to speak has become—even in an allegedly post-Freudian era—simply taken for granted.

Where did this narrative come from? To understand, we need to return to the history of hysteria at the end of the nineteenth century and recall the way it was conceptualized by people operating with assumptions drawn from "the power of suggestion." If suggestion and suggestion alone defined one's understanding of mind-body interactions, the choices were clear: either hysteria was organic, a product of pathological physiology and real, or it was psychological, a product of mere suggestion and therefore—as Bernheim's debunking of Charcot seemed to suggest—effectively fake.

But Freud offered another way to think about hysteria. For him it was true that the solution to hysteria lay in the mind; true, therefore, that Charcot had been wrong in his attempt to make hysteria map onto the assumptions of a too-rigid physicalist medicine. Nevertheless, it was also true that there was more going on in hysteria than could be accounted for by invoking the concept of suggestion. How did Freud know? Well, by listening to hysterics, rather than simply taking photographs of them or trying to manipulate them using hypnotic commands. When Freud listened to his hysterical patients, they themselves, he said, began to reveal what really lay behind their suffering and their symptoms. Meaningful symbolic connections began to emerge between their symptoms and various secret wishes, taboo thoughts, and traumatic memories they thought (or hoped) they had forgotten. Moreover, Freud also discovered that tracing those connections, revealing those troubling memories and thoughts, did not just explain hysteria: it also cured it. This was because, when patients' minds finally found the courage to speak aloud previously resisted or denied truths about their situation, their bodies no longer had to speak those truths for them in the form of physical symptoms. In other words, confessing the truth made them well.

This is a new narrative for the history of medicine, a product only of the last years of the nineteenth century. Nevertheless, it feels very plausible and persuasive; and this is because on a deeper level the narrative really isn't new at all. We live in a culture that has believed for a long time that secret sins can make people sick in both body and mind, and that conversely the open admission of sins can heal. The Christian tradition repeatedly emphasizes that the first step to healing lies in open confession: "Confess your faults one to another, and pray one for another, that ye may be healed" (James 5:16). In the medieval European tradition, possessed people generally could not be relieved of their affliction until the demon had confessed who or what lay behind the possessed person's plight. In this way, "the body that speaks" provides a secular variant on a much older Judeo-Christian understanding. In place of an older, religious healing ritual involving (say) the confession booth and the priest, "the body that speaks" offers a new medical ritual involving the psychoanalytic couch and the doctor.[2]

Freud as nerve doctor and hypnotist

Freud was able to think differently about hysteria than either Charcot or Bernheim in part because his relationship with patients was quite different from theirs. Freud was a "nerve doctor," someone who treated well-educated and often wealthy women suffering from hysteria and a variety of ill-defined somatic and emotional problems in his own private practice. Unlike a doctor in a big municipal or university research hospital, a nerve doctor needed to produce practical results for his patients—he had to make them feel better—or he wouldn't be practicing long. He also had to

Sigmund Freud (1856–1939), photograph by Wilhelm Enge, ca. 1885.

treat his patients with sufficient respect and tact to ensure that they did not take their business elsewhere. In this sense, Freud was dependent on his patients in a very different way than either Charcot or Bernheim.

Actually, Freud might well have preferred a career more like theirs. Reluctantly, he had accepted that his preferred career as a scientific researcher (his first professional publication was on the sexual organs of eels) was unlikely to succeed, in no small part because of institutionalized anti-Semitism in the Viennese university system. It had been a bitter pill to swallow, but Freud was eager to finally settle down with the girl who had "been waiting for me in a distant city for more than four years" (as he put it in his autobiography);[3] and he could not contemplate marriage in the absence of financial security. In 1886, therefore, he left the academy and opened a private medical practice in Vienna as a "specialist in nervous diseases." On September 13 of that year he married his fiancée, Martha Bernays.

In 1885, however, just before he made these irrevocable moves, Freud was successful in securing a travel bursary to spend three months at the Salpêtrière, where he attended Charcot's lectures on hysteria. The French neurologist impressed him enormously: "Charcot, who is one of the greatest physicians and a man whose common sense borders on genius," he wrote to his fiancée at this time, "is simply wrecking all my aims and opinions. I sometimes come out of his lectures as from Notre Dame, with an entirely new idea about perfection."[4] Nevertheless, even as Freud delighted in Charcot's brilliance, he couldn't ignore the challenge that was increasingly being mounted to the great man's reputation by Bernheim. Though he tried to defend Charcot to his colleagues for a while, over time Freud also became increasingly impressed with Bernheim's point of view.[5]

The reasons for this were partly practical. For a nerve doctor like Freud, Bernheim's ideas were easier to apply than Charcot's; one could do something with them in the consulting room, and with apparently far more hope of success than those that generally resulted from the conventional tricks of the nerve doctor trade, like massage and electrotherapy. By 1887 Freud had begun to experiment with Bernheim's method of therapeutic suggestions, and was reporting "all sorts of small but remarkable successes";[6] and a year later, he translated Bernheim's book on suggestion into German, declaring in a preface that "the use of hypnotic suggestion provides the physician with a powerful therapeutic method."[7]

In 1892, Freud published a case in the medical literature on his use of hypnotic suggestion to treat a woman who had a newborn infant and had developed simultaneously an aversion to all food and to nursing her baby:

> I made use of suggestion to contradict all her fears and the feelings on which those fears were based: "Have no fear! You will make an excellent nurse and the baby will thrive. Your stomach is perfectly quiet, your appetite is excellent, you are looking forward to your next meal," and suchlike.[8]

After several sessions, he reported the woman had made a full recovery and nursed her child for eight full months. Nevertheless, Freud complained, he had the strong sense that neither she nor the family properly appreciated his critical role in her cure. On the contrary, they seemed rather hostile toward him throughout the whole process. Much later, the woman herself finally indicated why: " 'I felt ashamed,' the woman said to me, 'that a thing like hypnosis should work where I myself, with all my will-power, was helpless.' I believe . . . neither she nor her husband have overcome their aversion toward hypnosis."[9]

Listening to Anna O.

Ambivalent and ungrateful patients were not Freud's only problem with hypnotic suggestion; the method also proved more finicky and less reliable than he had originally imagined. Thus, he had plenty of reasons for taking an interest in a novel and potentially more promising approach to hypnosis that had been used more than a decade earlier (from December 1880 to June 1882) by Josef Breuer, in his treatment of a young woman, Bertha Pappenheim, suffering from severe hysteria. Not only had Breuer's approach to hypnosis apparently worked; it also appeared to cast new light on the real causes of hysteria. Freud persuaded Breuer to write up his treatment of this young woman, and it was published as the case of "Anna O." in a coauthored 1895 text, *Studies on Hysteria*.

Anna O. is one of the most famous cases in the annals of psycho-

Bertha Pappenheim ("Anna O.") (1859–1936). © *Mary Evans Picture Library / Sigmund Freud Copyrights*

analysis. This is doubly ironic, since not only was this not one of Freud's own cases (he never even saw the patient), but it is now well established that the patient actually failed to recover under Breuer's care and had to be sent to a sanitarium in Switzerland (Breuer falsified the case study to hide that crucial fact).[10] It is true that Bertha Pappenheim did recover fully later on and became active in multiple social causes, founding an orphanage for Jewish girls and leading an international campaign against prostitution. However, she never credited either Breuer or psychoanalysis with her recovery, and always expressed great disdain for the entire enterprise.[11]

Nonetheless, the significance of Breuer's story about a character called Anna O. (as opposed to a real woman named Bertha Pappenheim) remains unaltered: this is the story, more than any other, that has taught us that the body can speak, and that what it speaks about are often precisely those things that the conscious mind would most wish left unsaid. Regardless of its shaky relationship to the biographical facts, therefore, the case of Anna O. demands our attention.

Breuer opened the story with an admiring description of its protagonist: Anna O. was "markedly intelligent," possessed of "great poetic and imaginative gifts" that were, nevertheless, tempered by a "sharp and critical common sense." She was, moreover, Breuer insisted, "*completely unsuggestible*," influenced, he continued, "only by arguments, never by mere assertions."[12] Notwithstanding all her gifts and virtues, she was severely hysterical, with symptoms that included disorders of vision, coughing, paralyses, and regular spells of somnambulism. The first symptoms had come on when she was nursing her father through a long illness, and by the time Breuer began visiting her a year later, she was almost wholly incapacitated and housebound.

This established, Breuer then explained the remarkable things that happened to Anna O. as he began to treat her using hypnosis. In a trance state, Anna O. was able to talk to Breuer about all the disturbing thoughts she had in her head; and the process of such mental "chimney sweeping" (as she herself termed it) seemed to reduce her symptoms and helped her to sleep. After a while, the chimney sweeping exercises became more focused. Anna O. began to recall a series of previously forgotten events that were causally linked to the appearance of her various symptoms. Even stranger, each time she recalled a forgotten event and experienced the feelings associated with it, the relevant symptom disappeared. For example:

> [D]uring a period of extreme heat . . . the patient was suffering very badly from thirst; for, without being able to account for it in any way, she suddenly found it impossible to drink. . . . one day during hypnosis she grumbled about her English lady-companion whom she did not care for, and went on to describe, with every sign of disgust, how she had once gone into that lady's room and how her little dog—horrid creature!—had drunk out of a glass there. The patient had said nothing, as she had wanted to be polite. After giving further energetic expression to the anger she had held back, she asked for something to drink, drank a large quantity of water without any difficulty. . . .[13]

The conclusion seemed clear: Anna O. was not suffering from degenerative heredity or localized damage to the nervous system, but from bad memories—memories with a strong emotional charge that she had tried to excise from her mind. Her case showed, however, that there was no hiding from such memories; in the end, each had found an alternative expression in bodily symptoms. And what was true for Anna O. was true for all hysterics, Breuer and Freud concluded. Hysteria was a disease caused by the repression of traumatic memories, memories whose emotional significance had not been properly acknowledged by the patient and therefore found expression through bodily symptoms.

How did one help such patients? The solution in Anna O.'s case became the general rule: one helped by encouraging the patient both to remember things and to acknowledge the feelings associated with these things. Each memory needed to be turned into words, into a true story that could be spoken aloud and shared so the body no longer needed to function as its mute narrator.

Listening, but not believing

Scholars of psychoanalysis who have studied this early work of Freud and Breuer have found much in it to admire. "Freud . . . changed the role of patients through their talk, by having them tell their tale in their own way."[14] Nevertheless, things quickly got more complicated. Within a few years, Freud indicated he was no longer satisfied with his initial understanding of hysteria. He had been wrong, he said, to go along with Breuer in thinking that hysteria was caused by repressed traumatic memories in general. His continuing clinical experience suggested instead a far more radical alternative: hysteria, he said, was caused exclusively by repressed traumatic memories of early childhood sexual abuse (he called it sexual seduction), generally by a parental figure and usually by a father.[15]

The seduction theory of hysteria was radical, all right, but was it true? Certainly, it didn't help Freud's confidence when he first presented the seduction theory at a scientific meeting in 1896 and found himself the target of abuse and ridicule. Within four years he had retreated, though for reasons he always insisted had to do with logic and evidence rather than professional pressure. The frequency with which patients reported sexual abuse, he said, had begun to make him suspicious: there just could not be that many incestuous fathers and uncles walking the streets of Vienna. So why did patients say what they did if their memories were in fact not trustworthy?

Freud's answer went like this: hysteria was indeed caused by traumatic sexual memories from childhood; that part of his original theory had not been wrong. However, patients were not remembering actual sexual encounters with parents; instead, they were remembering childhood fantasies—deeply shameful and denied for years—that they had had about such sexual encounters. These fantasies were the real secrets that the bodies of hysterical patients were betraying.[16] True crimes of abuse from childhood had little if anything to do with it, whatever the women themselves might say.

In the 1980s, this move from a trauma-based to a fantasy-based theory of hysteria would become a lightning rod for feminists and others who believed they had rediscovered the reality of widespread childhood sexual abuse. In those years, Freud was accused of betraying women who were telling him ugly truths that people didn't want to hear back in nineteenth-century Vienna.[17] For Freud himself, though, abandoning the seduction theory of hysteria would turn out to be enormously liberating. Suddenly, he no longer had to limit himself to investigating the etiology of a very specific disease called hysteria; he could instead ask questions about universal stages of psychosexual development in early childhood, the role of fantasy and desire in the development of neurosis, and much more. In this sense, he was free to develop the psychoanalysis we all know today.

All that is true and important. At the same time, Freud's turn from trauma to fantasy had a different consequence that should interest anyone concerned with the historical development of "the body that speaks" as a new kind of medical narrative. It was now clear, at least to Freud and his followers, that the patient was not going to be the best judge of what was going on with her body, that she should not necessarily be allowed the last word. From now on, the doctor's interpretations of the coded messages from the body—the meaning of a particular complex of symptoms or "choice" of disease—would emphatically trump that of the patient, and it would do so even if (sometimes especially if) the patient resisted or denied those interpretations.

That said, Freud's decisive rewriting of "the body that speaks" did not mean that his earlier version of the narrative vanished from the scene. It meant instead that there were now at least two possible ways to tell stories about speaking bodies: one that turned on the recovery of buried memories of actual events, and a second that turned on the recovery of childhood fantasies and feelings that were taboo. Freud had rejected the first way and insisted that his followers do likewise, but he could not control the activities and ideas of people outside his circle of students and disciples. In France, for example, the influential psychologist Pierre Janet also believed that the bodies of hysterics told tales—but continued to insist that the things about which they spoke were not just fantasies, sexual or otherwise, but real-life events so painful that a patient could not integrate them into his or her normal consciousness.[18] Moreover, he insisted that it was not true, as Freud said, that the only way to cure an hysterical body was through remembering. One could also achieve lasting cures, Janet felt, through a process of suggested *forgetting*, by using the power of suggestion to transform authentic but intolerably traumatic memories into false but inoffensive memories that would allow the patient to be at peace.[19]

At the same time as he challenged Freud on all these fronts, Janet began to say he was actually the originator of the theory of hysteria that found its source in traumatic memories and that Freud had borrowed all the key ideas for his own theories from him. The British psychoanalyst Ernest Jones later told the story of how, in 1913, he had defended Freud's originality in what he called a "duel between Janet and myself . . . at the International Congress of Medicine." According to Jones, Janet had backed down before Jones's onslaught, and that had effectively been the end of the matter.[20]

Listening to the shell-shocked body

Except that it really wasn't. Within less than two years, the now multiple ways of thinking about speaking bodies—as victims of trauma or secret concocters of fantasies, healed through psychoanalytic probing of repressed memories or healed through suggestion—would go head-to-head and do so in a context where the stakes could hardly have been higher. In 1914, World War I broke out in Europe. This was a war unlike any that either soldiers or military planners had ever known. It was the first war to make military use of airplanes and other flying machines; the first war to use tanks, submarines, and chemical gas in battle; and the first war to employ heavy artillery on the field.

It was also the first war to send thousands of young solders for months at a time into trenches that stretched across hundreds of miles, and into which shells and other kinds of artillery could land at any time. On the Western Front, stretching from the North Sea to the Swiss frontier, the trenches became notorious for their regular massive bombardments by heavy artillery, for the rats that lived in them and gorged themselves on corpses, and for the miles of barbed wire that soldiers had to cut through—often under fire—before the line could advance.

In this hellish environment, military physicians began encountering soldiers with a range of odd disabilities. Some had trouble seeing, hearing, smelling, tasting. Others had trouble walking or talking. Still others developed strange twitches, paralyses, and convulsions. And still others suffered from memory loss, while shaking uncontrollably all day long.

What all the men had in common was that they had not been obviously injured in any way, nor did any of them suffer from any evident organic disease. The term "shell shock" was invented to designate these cases, derived from a theory proposed by the English physician Charles Myers that all these symptoms might have their source in invisible shocks—what were called "traumas"—to the nervous system brought about by proximity to artillery explosions.[21] When, however, the numbers of patients with strange physical symptoms began to include soldiers who had never been close to artillery and, indeed, had yet to even see combat, it was clear that a different kind of theory was needed.

One explanation ready at hand, of course, was that what was wrong with these men was precisely the same thing that afflicted malingerers everywhere—namely, nothing. What these men with their strange symptoms really suffered from, insisted some, was not a medical condition at all but a simple loss of nerve, a shameful cowardice. In the words of one British officer at the time, "shell shock must be looked upon as a form of disgrace to the soldier. . . . Officers must be taught much more about man mastership in the same way as horse mastership. . . . It is all to a great extent a question of discipline and drill."[22] In some camps, shaming, taunting, and the infliction of pain were used to goad soldiers back to fighting fitness. In other instances, the ultimate deterrent was employed: court martial and death by firing squad.

Some of the physicians who had seen these men, however, were not persuaded that matters were so simple. For them, it was clear that, even if the symptoms suffered by these unfortunate young men did not have any physiological cause, they nevertheless had a real cause—a mental cause. People were reluctant to use the word "hysteria"—still associated with women and weakness—but it is obvious from the writings of the time that everyone realized that this is what they were facing: an epidemic of male hysteria.

As discussed in the previous chapter, some people proposed to use hypnotic suggestion to cure these young men—taking advantage of the fact that soldiers were accustomed to doing what they were told. Moreover, not everyone thought actual hypnosis was necessary to get results—emphatic suggestion on its own could be just as effective. In Germany, it was the charismatic medical hypnotist Max Nonne, a man who had studied earlier with both Charcot and Bernheim, who became most closely identified with the use of suggestive therapy for the treatment of war neurosis.[23] In the English-speaking world, the Canadian physician Lewis Yealland took the lead in developing a version of this approach: the military doctor, he said, must break down the patient's attachment to his fixed idea of injury by brooking no contradiction. Yealland would thus tell soldiers they were going to recover and would not be allowed to leave the room until they did. He shamed them by insisting that they needed to act like men, like heroes, and pull themselves together, repeatedly shocking the malfunctioning parts of the patients' bodies with a strong faradic electric current as he spoke. Doctors like Yealland were not interested in probing the psychological meaning that any of these bodily symptoms might have for the soldiers. What mattered to them was that as many of these men be made fit for fighting again as quickly as possible.[24]

Yet even though men like Yealland and Nonne claimed good results, others were less sure. Even leaving aside any moral misgivings some had about the harshness of a military-minded suggestion therapy, critics began to note that many of the soldiers treated using suggestion either relapsed after being released or developed new symptoms once they were back in combat. The power of suggestion, some people began to say, could perhaps cure symptoms, but this was very different from the challenge of curing the disorder itself, curing the patient.

In England, one of the people who began to talk this way was the psychiatrist and anthropologist W. H. R. Rivers, who directed a private hospital for shell-shocked soldiers in Scotland called Craiglockhart. Rivers was one of a minority of physicians at the time who believed that the therapeutic ideas of Sigmund Freud offered one of the most promising approaches available for shell-shocked patients. While his wartime experience had made him distinctly skeptical of Freud's view that the deep roots of hysteria lay in repressed sexual wishes and conflicts, Rivers insisted all the same that "there is hardly a case which this theory does not help us the better to understand—not a day of clinical experience in which Freud's theory may not be of direct practical use in diagnosis and treatment."[25] He spoke of shell-shock cases in which patients exerted tremendous amounts of energy in the effort to repress painful memories, while suffering at the same time from a whole series of distressing somatic symptoms. These symptoms could be alleviated best, he learned, by encouraging soldiers to confront the frightening memories and make some kind of peace with them.[26] In other words, confession of repressed memories proved to be healing, just as Freud had said, even when such memories had nothing to do with sex.

By 1919, Freud himself was almost conceding the same. The phenomenon of shell shock had indeed complicated his insistence that all hysteria is caused by childhood sexual fantasies, he admitted. It was now clear, he said, that one needed to distinguish between neuroses of war and neuroses of peace.

> [In] traumatic and war neuroses the human ego is defending itself from a danger which threatens it from without. . . . In the transference neuroses of peace the enemy from which the ego is defending itself is actually the libido, whose demands seem to it to be menacing. In both cases the ego is afraid of being damaged—in the latter case by the libido and in the former by external violence.[27]

By the time the war ended, the Freudian perspective on speaking bodies was no longer—at least in the eyes of some—just about sex and fantasy.

Bodies that speak with a German accent

And it would soon no longer be just about the hysterical body either. By the 1920s, Freud himself had become less interested in hysteria and more interested in the unconscious roots of more transparently "mental" disorders like anxiety and obsession. In those same years, however, some of his followers pushed debate in the other direction. They asked: had the master actually been radical enough in his treatment of the body? He had demonstrated the power of listening to hysterical bodies for the repressed messages they were telling, but what about bodies afflicted with other disorders—even disorders with a supposed somatic basis? Was it possible that the symptoms associated with these disorders also had a tale to tell, also represented some coded message from the unconscious?

There was prima facie reason to believe that some might. After all, some of Freud's followers pointed out, a number of chronic disorders like asthma, headaches, and ulcers have a tendency to wax and wane in intensity in ways that might be psychoanalytically meaningful. Why should this particular patient get an attack of asthma just before his mother-in-law visits? Why should that other patient develop debilitating headaches every time her spouse returns from one of his business trips?

The term that some began to use to describe disorders of this sort—disorders that seemed to be more than hysterical but less than fully organic—was "psychosomatic." By the 1920s, research and theory into such disorders were sufficiently developed, people felt, for the field to warrant its own name: "psychosomatic medicine." The word "psychosomatic" was not new; it had its origins in early-nineteenth-century romantic philosophy and literature, where it referred broadly to interactions between pathologies of the mind and body.[28] What was new was the decision by some of these early followers of Freud to identify the word broadly with a psychoanalytic agenda.

As an institutionalized program of theory, research, and practice, psychosomatic medicine developed in two distinct registers: an earlier German-speaking one and a later English-speaking, American-accented one. In the German-speaking countries (Germany, Austria, Switzerland), psychosomatic medicine institutionalized itself in the 1920s in ways that kept it closely allied with the professions of nerve doctoring and neurology (one of the key vehicles for the new ideas was the journal *Der Nervenarzt,* founded in 1928). At the same time, the field increasingly identified its cause with larger efforts in German-speaking European medicine to challenge mechanistic, dualistic, and reductionistic thinking and to turn medical practice in general in more holistic and naturopathic directions. Medicine was in crisis, these critics said, because doctors cared only about the data of the microscope and dissecting laboratory and had forgotten the suffering person. German psychosomaticists—armed with the insights of psychoanalysis, and frequently imbued with the spirit of such Romantic-era thinkers as Carl Gustav Carus, Johann Wolfgang Goethe, and Friedrich Wilhelm Joseph Schelling—believed they had an important part to play in helping medicine make more room for the whole person.[29] "From the first," as one of these men, Georg Groddeck, later recalled, "I rejected a separation of bodily and mental illnesses, tried to treat the individual patient [and] . . . attempted to find a way into the unexplored and inaccessible regions."[30]

Groddeck was a nerve doctor, director of a private sanatorium in Baden-Baden, and someone who often treated patients who came to his clinic as a last resort after trying many other types of treatment. The question, he later said, that he started to ask is: Why did the other treatments not work? Why did these people remain ill?

Groddeck found his answers not in the modern laboratory or modern textbooks written by physiologists and bacteriologists, but in principles of healing he had learned first from his former mentor, Ernest Schweninger. Schweninger was a naturopathic physician who believed that doctors never cure patients, but simply help remove barriers that keep patients from curing themselves. What were these barriers to self-cure? According to Groddeck, they had to do with resistance to the demands of a great unconscious force that exists within every person—a force so mysterious that Groddeck refused to give it a name, instead merely calling it "the It" (*das Es*). When people deny the demands of their "It"—and in this sense deny their own authentic needs—"the It" still insists on speaking its mind through the language of illness. For Groddeck, then, all illnesses, however severe, are best understood not as products of infection or defective heredity but as "drama(s) staged by the It, by means of which it announces what it could not say with the tongue."[31]

One of Groddeck's cases, first published in 1923, will serve to show how all this looked in practice. A married mother of two had come to his sanatorium suffering from severe pain in her hand. It turned out that some time back she had cut her right index finger while opening a can of fruit. The wound had become infected and she had developed a pus-filled abscess that was removed by the doctor. Instead of healing, the infection had spread, and eventually the first joint of the finger had to be amputated. But that didn't stop matters, because the trouble spread to the next joint, and she developed a necrosis, or localized death of the tissue, in that area. Things went from bad to worse, and finally the whole finger was removed and a skin graft carried out. At that point, the woman developed terrible pains where the nerves had been severed. These pains spread up her entire arm. Operations were attempted; narcotics were offered; but nothing helped.

On hearing the story of this woman's medical woes, Groddeck concluded that she was a person with a deep "resistance to becoming well." He launched three months of intensive psychoanalytic-style investigation, and this gradually produced a second story, the story behind the medical symptoms. It emerged that the woman was a Christian who—in a time when anti-Semitic sentiments in Weimar Germany were on the rise—had married a Jewish man. They had two children, but for economic reasons the husband had proposed that they should have no more and the wife agreed. Groddeck felt reasonably sure that, even though the decision had been his, the husband had experienced considerable ambivalence about it, since it is a religious duty in the Jewish tradition to have many children. His conflicted attitude had manifested itself in periodic impotence during the masturbatory sessions with his wife that had now replaced intercourse. When this happened, the wife was forced, much to her distaste, to assist her husband by rubbing his penis. To do so, she had to use "the thumb and index finger of her right hand"—the same digits that were now so disabled by injury and pain.

Groddeck did not stop the analysis here. He deduced that the woman's real reason for avoiding intercourse with her husband was not economic (as Groddeck pointedly noted, "she was a Christian," and thus not subject to the miserly impulses common to her husband's race) but emotional. She had done so, he suggested, "because she had gradually come to feel a repulsion against the Jewish race." She had repressed her conflicted feelings as long as she could, but in due course her unconscious had orchestrated an accident designed to expose the true conflict in all its details. The woman's only chance for recovery, Groddeck concluded, lay in acknowledging her inner conflict: on the one hand, her desire for more children; on the other, her revulsion toward her husband. She also needed, he said, to make her peace with the half-Jewish legacy of her children. We do not learn from this case history whether in the end the woman accomplished this. As written, the case stands as an unwitting testimony, not only to her personal woes but to a troubled and troubling time in Germany's social history.

Groddeck originally developed his ideas about illness as a symbolic language of the unconscious independently of Freud. In 1911, however, he happened upon some of Freud's writings and was reluctantly forced to concede that he, Groddeck, had been thinking "psychoanalytically" for some time without realizing it. Freud was happy to add him to his "flock of disciples."[32] By the 1920s, Groddeck's term *das Es* had even found its way into some of Freud's writings, where it was later translated into English as "the Id."

Freud may have claimed Groddeck, but he never really tamed him—and Groddeck knew it. As Groddeck cheerfully observed in his *Book of the It*, "everything in [here] that sounds reasonable, or perhaps only a little strange, is derived from Professor Freud of Vienna and his colleagues; whatever is quite mad, I claim as my own spiritual property."[33] Later, other German psychosomaticists would complain that Groddeck was irresponsibly speculative in his style and claims and not really representative of what they were all about.[34]

More respectable, in their eyes, was the internist Viktor von Weizsäcker, who directed a psychosomatic medical clinic in Heidelberg. Like all German psychosomaticists, Weizsäcker recognized a significant debt to Freud. As he put it in an oft-quoted sentence, "Psychosomatics will be psychoanalytic or will not be at all."[35] Nevertheless, he also acknowledged equally significant allegiances to Gestalt psychology, clinical neurology, and the existentialist philosophies of both Martin Heidegger and the Jewish theologian Martin Buber (a close friend).[36]

Weaving all these influences together, Weizsäcker's psychosomatic medicine took as its starting point the principle that the patient should be at the center of every clinical encounter, not as an *object* of medical investigation but as an experiencing *subject*. When that was done, it became clear that every illness was more than just a biological event; it was also a biographical event rich with (usually hidden) meaning—meaning that in turn further shaped the bodily experience. It was the physician's responsibility to help the patient recognize not just what was wrong with him but what his ailment meant to him. For this reason, Weizsäcker recommended opening a diagnostic interview by asking the patient, not "Where does it hurt?" or "When did it start?" but rather "Why this symptom? Why now?"[37]

And the answer, on some level, was almost always the same: the patient faced a dilemma or life choice that he or she did not know how to resolve, and illness was a way of forcing matters to some sort of resolution, often in a direction that the patient unconsciously wanted or needed, but could not easily acknowledge. In Weizsäcker's words:

> A situation is given, a trend develops, tension increases, a crisis comes to a head, there follows an outbreak of illness and with that, after that, the decision is made: a new situation is created and settles down; the profits and losses [from before] can now be disregarded. The whole functions like an historical unit: turning point, critical disruption, transformation.[38]

The "pathosophies," or illustrative case histories, that Weiszäcker collected in support of his views often had an almost aphorismic sensibility reminiscent of the Jewish-Chassidic folktale used to such good effect by his friend and colleague Martin Buber.

> An older spinster marries a railway officer who is also already advanced in years, but she still hopes for full marital happiness from this union. However, he is transferred to a distant city, and she cannot make up her mind to leave her mother and follow him there. Finally, after much negotiating back and forth, she does go to him. On the day of her arrival, he experiences a serious angina with complications, and after a long convalescence is disabled and pensioned. He then follows his wife back to her mother, and all live together now in the same place.[39]

All this implied an agenda for psychosomatics of organic disease considerably more ambitious than Freud would have countenanced, and Weizsäcker knew it. In a 1927 essay, he told the famous story of how the French neurologist Jean-Martin Charcot had privately admitted to a young Freud that, in cases of hysteria, there was "always" a sexual issue at stake. In his autobiographical account, Freud recalled how he had said to himself: "If he knows that, why doesn't he say so?" Weizsäcker then described an encounter he had himself had with an aging Freud, many years later, in which the great man had admitted that the sudden intrusion of an accident or organic disease in the life of a patient can cure that patient of his or her neurosis. Weizsäcker had then said to *himself*: "If he knows that, why doesn't he say so?" Obviously, Weizsäcker believed that it was up to him to "say so"—to develop the full clinical implications of the insight. In his 1927 essay, he concluded: "It seems as if, in science, there is almost a law whereby one epoch only says the one thing, while remaining silent about the other thing, that it also knows."[40]

What happened to this first flowering of psychosomatic medicine in Germany in 1933, with the rise to power of National Socialism? For many years, it was generally assumed that the entire enterprise had ground to an abrupt halt as the Jewish psychoanalysts fled and everything associated with Freud was brutally suppressed by the new regime. Thus, after 1945, men like Alexander Mitscherlich—successor to Weizsäcker in Heidelberg—came out and insisted that it was now up to his generation to reclaim and rebuild those repressed and victimized traditions in ways that would serve humane, democratic values.[41] And in this way, the German tradition of psychosomatic medicine would begin anew in 1945.[42]

More recently, however, it has become clear that matters in fact were more complex than some people had long wanted to admit. It is true that some key psychosomatic theorists (Georg Groddeck, Felix Deutsch) did leave Germany, either voluntarily or under duress, when the National Socialists took power.[43] Nevertheless, the rise of Nazism did not mark an abrupt end either to Freudian thinking or to psychosomatics in Germany. In Berlin, Matthias Goering—an Adlerian psychotherapist and (very usefully) a cousin of Reichsmarschall Hermann Goering—directed the German Institute for Psychological Research and Psychotherapy, where Freudian-inflected psychotherapies (stripped of all explicit reference to Freud) continued to be taught and practiced, and where Goering personally maintained a lively research interest in psychosomatic matters.[44]

At the same time, in Heidelberg, Viktor von Weizsäcker was able to continue his pioneering psychosomatic work apparently undisturbed and even was permitted to keep his private library of Freud's works at a time when they had been officially banned.[45] Having early on expressed support for Nazi medical euthanasia programs ("it would not be fair if the German physician thought that he need not make a responsible contribution to a politics of necessary extermination"),[46] Weizsäcker was, by the late 1930s, contributing a regular monthly column on psychosomatic matters to a journal of naturopathic medicine called *Hippokrates*.[47] This was no ordinary journal, but the official mouthpiece (since 1934) for a party-led effort to counter mechanism and materialism in medicine (assumed to have been largely caused by pernicious Jewish influences) and to create a "New German Medicine" (*Neue Deutsche Heilkunde*) infused with folk wisdom and holistic Hippocratic understandings of disease and healing.[48]

From the beginning, articles in *Hippokrates* had made clear that psychosomatic medicine would work alongside homeopathy and herbalism in the larger effort to create a more natural and earthbound approach to healing. The introduction of a new monthly column by a figure as distinguished as Weizsäcker, however, significantly raised the potential importance of the enterprise for New German Medicine as a whole. We do not know what Weizsäcker thought of the New German Medicine project, or what pressure, if any, he came under by the Nazi authorities to contribute that column.[49]

In the end, perhaps it hardly matters, since the attempt to transform German medicine into a psychosomatic, holistic enterprise largely failed. As Germany geared up for war, more hard-nosed, "mechanistic" medical leaders within the party and the SS—men with a range of practical technologies and concerns (racial screening, sterilization, military medical research, and ultimately methods of mass "euthanasia")—increasingly came to prevail over the "holists."

It is true that, by the last years of the war, German psychiatrists were again confronting large numbers of soldiers suffering from the paralyses, shaking, and other physical symptoms of battle trauma. This time, though, few if any had patience for the idea that the bodies of these soldiers might be trying to tell tales to which they should be listening; instead, discipline, electric shock, and sedatives were the preferred treatments of the day.[50] In this sense, German holistic psychosomatic thinking in the end really was sacrificed to the different imperatives of the Nazi war machine.

The American approach to speaking bodies

In the United States, things unfolded very differently. Where German psychosomatic medicine was part of a larger, increasingly politicized effort to bring wholeness and soul back into medicine, American psychosomatic medicine aspired above all to mainstream medical respectability. It was possible, its advocates said, to believe that the symptoms of illness "meant" something, that bodies could speak—but to do so while still believing in statistics, modern clinical research, and laboratory physiology.

What this meant in practice was that American psychosomatic medicine focused less on holistic concepts of mind-body unity and more on exploring the effects of particular psychological factors on particular diseases. Since the late nineteenth century, modern medicine in general had been committed to the doctrine of specificity: it had insisted that there are specific diseases (people don't just get sick; they get sick with *something*), and that every specific disease has a specific cause (people who get infected by tubercular bacteria always get tuberculosis; they don't get scarlet fever). American psychosomatic medicine also believed in specificity. It was just that, in its case, it aimed to identify the specific causal pathways or relationships that connected body and mind, biology and biography.

The American psychosomatic pioneer Flanders Dunbar is a case in point. Dunbar's role in the rise of American psychosomatic medicine was wide-ranging: it was she who, through her mammoth 1935 book *Emotions and Bodily Change,* helped produce consensus within professional medical circles that the mind-body connection warranted further study;[51] she who, in collaboration with psychoanalytic émigré Franz Alexander helped found the flagship journal of the American Psychosomatic Society, *Psychosomatic Medicine*, and served on its editorial board for eight years.[52]

Dunbar is best remembered, however, for her controversial efforts to link specific diseases to specific personality profiles. Working at New York–Presbyterian Hospital, Dunbar analyzed interviews with some thirteen hundred patients suffering from a range of chronic disorders. She concluded that specific kinds of character traits—especially the varying ability of patients to inhibit or express emotions—were associated with susceptibility to specific kinds of disorders. The evidence, she felt, was plain to see in the interview records.

> The *accident-prone* patient says, "I always have to keep working. I can't stand around doing nothing. When I get mad, I don't say anything. I keep it in and do something."
>
> The *hypertensive* patient says, "I always have to say 'yes.' I don't know why. I am always furious afterwards"; or "I'm angry but I never like to fight. I don't know why. Something must have happened once." "Argument is my long suit. I could argue all day long."
>
> The *asthmatic* patient says, "Doctor, it's terrible; I don't know what I might do. I'm constantly on the verge of killing somebody or injuring myself; *you've* got to keep me from it, I'm not responsible for myself."
>
> The *arthritic* patient says, "Everything I do hurts, but I have to keep on moving."[53]

During World War II, the American military took an interest in Dunbar's analysis of character types, hoping it might help officers avoid assigning soldiers to posts in which they were unlikely to do well.[54] In the popular press, it was her work specifically on so-called accident-proneness (a term she coined) that had the most resonance. As late as August 1969, an article in *Family Weekly* highlighted Dunbar's work in the context of a call for more attention to be given to accidents as "one of the nation's major health problems." Chronic repression of anger and guilt, the article intoned, "can be as dangerous as faulty brakes or a broken step." Becoming aware of these "human factors," however—and seeking psychiatric help if necessary—could help more people avoid dangerous mishaps.[55]

How far were those committed to this perspective prepared to go? If chronic suppression of negative emotions could cause us to wheeze, develop high blood pressure, and crash our automobiles on the highways, could it also cause us to grow tumors? Did there exist, in other words, not just an "accident-prone" personality but a "cancer-prone" personality?

Many thought so. As early as 1940, W. H. Auden had captured the emerging psychoanalytic consensus on what such a person might be like. His darkly comic poem "Miss Gee" tells the story of a spinster who keeps her "clothes buttoned up to the neck," knits for the church bazaar, rides a bicycle with a "back-pedal," and is vexed by unwanted dreams of passion. Finally she begins to feel unwell.

She bicycled down to the doctor,
And rang the surgery bell;
"O, doctor, I've a pain inside me,
And I don't feel very well."

Doctor Thomas looked her over,
And then he looked some more;
Walked over to his wash-basin,
Said, "Why didn't you come before?"

Doctor Thomas sat over his dinner.
Though his wife was waiting to ring,
Rolling his bread into pellets;
Said, "Cancer's a funny thing.

"Nobody knows what the cause is,
Though some pretend they do;
It's like some hidden assassin
Waiting to strike at you

"Childless women get it,
And men when they retire;
It's as if there had to be some outlet
For their foiled creative fire"[56]

Both aggressive feelings and sexual urges, psychiatrists and the general public increasingly agreed, could only be denied for so long; if a person buried them over the course of a lifetime, eventually they turned into tumors.[57] In 1960, when the writer Norman Mailer was arraigned for stabbing his wife in the chest and abdomen with a penknife, his defense lawyers explained that their client believed that, had he repressed his rage at that moment, he would have gone on to develop cancer.[58] In the end, his wife decided not to press charges, so we do not know how the court would have judged the defense, but the fact that it could have been seriously offered is telling in itself.

Dunbar's focus on personality type helped set one course for American psychosomatic medicine, and the work of her European colleague Franz Alexander—a psychoanalytic émigré who arrived in the United States in the 1920s—was critical in setting another. For psychosomatic medicine to succeed as a field, Alexander insisted (nudged on by some insistent officers from the Rockefeller Foundation, who were funding his work), Freudian thinking needed to be integrated with the best thinking of the time on the physiology of emotions, especially the exciting new work being done on "fight or flight" by Walter B. Cannon at Harvard.[59] Alexander's perspective on the genesis of psychosomatic disorders thus came to consist of two claims: (1) there is a specific relationship between chronic repressed emotional conflicts and specific diseases; but (2) disease itself is not caused directly by the repressed emotions, but rather by the fact that chronic repression of different specific emotions has the effect of chronically stimulating or activating different specific vegetative organs in one's body—the heart, lungs, circulation, gut, and more—until they finally begin to malfunction.

In the end, Alexander and his school concluded that the research pointed to the existence of seven kinds of psychosomatic disorders resulting from different kinds of chronic excitation: bronchial asthma, peptic ulcer, ulcerative colitis, thyrotoxicosis, essential hypertension, rheumatoid arthritis, and neurodermatitis. Each one of these diseases was associated with its own sad story. Ulcers were said to be caused by unresolved dependency issues and longing for love (and affected the stomach and gut because patients unconsciously associated such feelings with nursing and feeding). Asthma was said to be caused by unresolved dependency issues (the wheezing was supposed to be an unconscious infantile cry for the mother). Hypertension was caused by unconscious feelings of aggression, usually of an Oedipal nature. And so it went.[60]

By and large, American mainstream medicine proved cautiously open

Schematic representation of specificity in the etiology of peptic ulcer

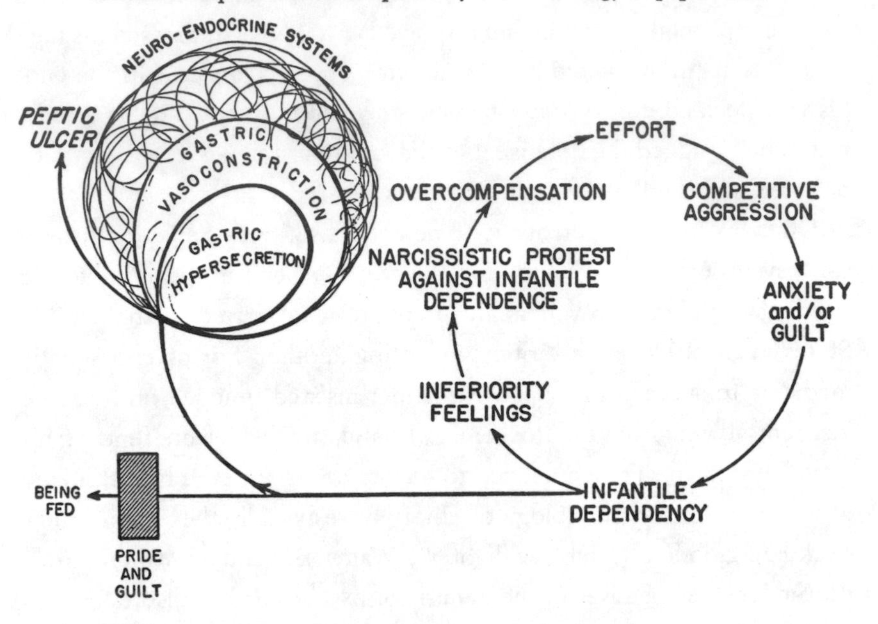

Franz Alexander's 1950 textbook representation of the ways in which specific repressed emotions with roots in life history factors may interact with physiological systems in the body to produce a peptic ulcer. From *Psychosomatic Medicine* by Franz Alexander. *Copyright 1950, 1987 by W. W. Norton & Company, Inc. Used by permission of W. W. Norton & Company, Inc.*

to Alexander's approach to psychosomatics. Politically, he had done several things right. He had publicly distanced his school from the speculative and expansive romantic approaches of European colleagues like Groddeck, insisting instead that psychosomatic medicine needed to develop its own vision of disease specificity that would be as rigorous as that of somatic medicine. He had insisted too that psychoanalysis on its own was not enough; Freudian methods needed to be integrated with data from the physiology laboratory. Perhaps most important of all, he had clearly signaled that he was not interested in colonizing all of medicine on behalf of Freud but simply wanted to claim a cluster of chronic disorders that mainstream medicine had long been notoriously unsuccessful in treating anyway. Why should anyone mind letting the Freudians take a crack at them?

By the 1950s, Alexander's textbook *Psychosomatic Medicine* had become the standard American reference of its time, notable not only for its particular arguments but also for its striking visual diagrams—more akin to something one would expect to see in a biochemistry textbook. Meanwhile, other colleagues were busy confirming this or that claim for psychosomatic specificity in large-scale studies. A 1957 study of ulcers in Army recruits was particularly influential (it turned out those soldiers really did have excessive infantile desires for love and feeding).[61] Discussions of the Alexander approach to psychosomatic medicine also began to appear in the popular press, from *Reader's Digest* to *Ladies' Home Journal*. And in 1950, this approach had a kind of milestone moment in terms of its penetration into the wider culture: it became the basis of a hit number, "Adelaide's Lament," in Frank Loesser's Broadway musical *Guys and Dolls*. Adelaide is a cabaret dancer of a certain age who has been unable, after thirteen years, to persuade her ne'er-do-well gambler fiancé to marry her. She also suffers from chronic respiratory problems, for which she seeks medical help. Her doctor suspects a psychosomatic cause, and lends her a textbook (Alexander's, perhaps?) to help her better understand her own case. She opens it up and begins to read:

It says here . . .
"The average unmarried female
Basic'lly insecure
Due to some long frustration
May react
With psychosomatic symptoms
Difficult to endure
Affecting the upper respiratory tract."
In other words,
Just from waiting around
For that plain little band of gold
A person can develop a cold![62]

The most important thing to appreciate about "Adelaide's Lament" is not that it was poking fun at the ideas of American psychosomatic medicine. What is important, rather, is that no one needed to have the joke explained to them. In other words, by 1950 the idea that bodily symptoms of illness can be a way in which the unconscious makes known its suppressed emotions and desires had become as much part of the popular culture as the Freudian ideas of penis envy and sibling rivalry. Like psychoanalysis itself, psychosomatic medicine had come of age.

Psychosomatic medicine goes alternative

The moment would not last. By the 1970s, psychosomatic medicine as a field was in trouble; by the end of the 1980s, most people felt it was in a state of critical decline, a victim of a combination of resurgent reductionism in biomedicine, a loss of focus within the field, and a general decline in the fortunes of Freud.[63] In 1988, the president of the American Psychosomatic Society summed up his sense of the state of the profession in strikingly gloomy terms:

> [I]n an organization committed to an increasingly nebulous and ambiguous concept called "psychosomatic medicine," who are we? . . . [W]e are essentially invisible as a society to the general public and also to most of the medical community. . . . [W]e exist in partial vacuum and in isolation from the vast majority of our colleagues in medicine, psychiatry, and psychology. . . . [T]he average age of our membership is 50 years, or, put another way, about half our members are eligible to join the American Association of Retired Persons.[64]

Today, almost twenty years later, the journal *Psychosomatic Medicine* still exists, but Freud and psychoanalytic thinking are almost nowhere present in its pages. As a journal, it must make its way in a world in which ulcers are now understood to be caused not by unresolved longings for mother love but by bacteria (the man who discovered the connection, Barry Marshall, won a Nobel Prize for this finding in 2005).[65] It must also make its way in a world where asthma is a disease caused by allergens, not anxieties. The website of the U.S. Environmental Protection Agency offers a quiz designed, among other things, to disabuse people of various false beliefs about asthma. The first question reads: "True or False?

Asthma is an emotional or psychological illness," to which the correct answer is: "False! Asthma is not an emotional or psychological disease, although strong emotions can sometimes make asthma worse."[66]

Nevertheless, to acknowledge a decline in mainstream acceptance of psychosomatic medicine is very different from saying that its key narrative—the idea of the speaking body—no longer holds power in our culture. On the contrary, the speaking body continues to hold considerable power; but it does so in the busy, restless margins of health care, and particularly in the worlds of alternative, feminist, and holistic medicine. There it functions above all as a call to personal empowerment: what we learn from the idea of the speaking body, we are told, is that patients need to trust their bodies and the messages their bodies send, whatever doctors and their fancy machines might say.

To see how this shift has occurred, consider, for example, the recent history of the idea of trauma. In the early 1980s, the American Psychiatric Association responded to intense lobbying by Vietnam veterans by recognizing a new diagnostic category: post-traumatic stress disorder (PTSD). This diagnosis replaced the diagnostic category "war neurosis," which during World War II had in its own right replaced the World War I concept of shell shock. Unlike shell shock, war neurosis (sometimes "battle fatigue") had adhered relatively strictly to later psychoanalytic orthodoxy by insisting that, while every man had his breaking point, those who broke down sooner rather than later did so as a result of preexisting neurosis. The idea of post-traumatic stress disorder was closer to the old idea of shell shock: war itself is a trauma, its advocates insisted, that results in breakdown and dysfunction.

That was not the end of the matter, however. The 1980s saw an astonishingly rapid expansion of the PTSD diagnostic category, much of it driven now by a new wave of feminist activism. In these years, the spotlight was put on the widespread realities of domestic violence, rape, and (perhaps most radically) the rediscovery of the existence of sexual abuse of young children. By the end of the 1980s, battered wives and abused children had been set alongside traumatized soldiers as potential candidates for a diagnosis of PTSD. In her 1989 best-selling book *Trauma and Recovery*, the activist psychiatrist Judith Lewis Herman made the links explicit:

> [This] is a book about commonalities: between rape survivors and combat veterans, between battered women and political prisoners, between the survivors of vast concentration camps created by tyrants who rule nations and the survivors of small, hidden concentration camps created by tyrants who rule their homes.[67]

The expansion of the PTSD category beyond the battlefield and into the home had the effect of creating a substantial cottage industry of self-help books and psychotherapies designed for the thousands of people interested in tracing their chronic neuroses in adulthood to unmastered memories of childhood sexual abuse. The thing was, many of these alleged victims had no clear memories of such abuse. Could they still qualify as traumatized? The answer was yes, because even if their minds did not consciously or properly remember, their bodies most assuredly did. Did a person suffer from colitis? From arthritis? From otherwise unexplained headaches? From something still worse? The persistence of disorders like these in adulthood, people said, often functioned as the body's way of bearing witness to suffering first experienced in childhood but never properly acknowledged and processed. This was because, in contrast to the often self-deceiving, self-denying mind, "the body keeps the score," "the body never lies," "the body remembers."[68]

So, one way in which the speaking body has survived in our time has been through its recruitment to the quasi-political task of vouching for the reality of childhood trauma. Another has been through arguments that women in particular are prone to living in inauthentic ways that deny their real needs. The body alone, it is said, betrays these uncomfortable facts in the language of meaningful, but often debilitating, chronic medical conditions. Why, for example, do so many women today suffer from infertility or other reproductive problems? According to Christiane Northrup, a physician and popular writer on women's health and a third-wave American feminist, more often than not such problems are the body's way "of expressing the psychological and emotional wounding it has experienced from a culture that is often not supportive of women and their feminine values."[69] The Freudian psychosomatic literature from the 1940s and 1950s is critical to her case:

> Women who habitually miscarry or who have an incompetent cervix sometimes also have difficulty accepting motherhood and their feminine role. . . . They frequently choose dependent, nonverbal husbands and have restricted social outlets and low adaptability. Due to their aloofness, they are often unable to take part in life around them. . . . Another study found that "habitual aborters" . . . [feel] guilty about directly expressing their anger at other people's demands, [so] their frustration builds until their body responds with a physical illness. Miscarrying the child . . . relieves the tension that has built up in their bodies.[70]

Similarly, since the 1980s, some in the holistic health community have asked why are so many of us apparently vulnerable to cancer? Why do so many fail to recover? The answer they also offer is that too many people have failed to live emotionally honest lives. Cancer is a wake-up call from the body that makes clear just how steep is the price for all these too common failures of personality and courage.

The emergence of this way of talking about cancer in holistic health circles might seem a surprising one. In the 1970s, the prominent cultural critic Susan Sontag developed cancer herself and was horrified to discover that the old psychoanalytic belief in cancer as a disease of emotionally repressed people was still in wide public circulation. This was just ignorant talk, she snapped. People also used to believe there was a relationship between certain personality types and tuberculosis, until modern medicine discovered the bacterium actually responsible for the disease. It was no different with cancer, even if we did not yet know exactly what the specific material cause for it was. In her influential book *Illness as Metaphor*, Sontag wrote flatly: "Illness is *not* a metaphor . . . the most truthful way of regarding illness—and the healthiest way of being ill—is one most purified of, most resistant to, metaphoric thinking."[71]

This was the era in which the United States believed it was engaged in a "war on cancer," so the odds may then have looked pretty good that, in the next years, psychosomatic approaches to cancer would indeed go the way of previous psychosomatic or metaphorical approaches to tuberculosis. But things did not happen quite that way. Some progress was made on certain specific forms of cancer like childhood leukemia; but by the 1980s, the absolute numbers of cancer were on the rise, with a particularly worrying spike in cases of breast cancer. At the same time, new forms of aggressive chemical and radiation therapy were at best only partially successful; they rarely cured. In fact, by the 1980s some people began to say that cancer had become an experience in which the treatments almost kill you, and then you die anyway.

The growing discontent with mainstream medical approaches to cancer at this time helped create a highly receptive market for various alternative approaches to this disease that emphasized both gentleness and patient empowerment. In this context, there began to be renewed interest in the old psychoanalytic idea that emotionally repressed personalities are more susceptible to cancer than other people; but the idea was given a new twist. Now it was no longer a question of consigning all the lonely, prim, and proper Miss Gees to their inevitable fate, but of giving a new kind of hope to people who felt abandoned by mainstream high-tech medicine.[72] In the words of Lydia Temoshok, one of the psychologists who helped fuel new interest in this area, "The discovery that emotions can influence cancer risk and recovery means that we all have potentially powerful allies against cancer: our own hearts and minds."[73]

Of all who argued this way, none was more influential than the Yale University surgeon Bernie Siegel. As Siegel later told the story, it was his cancer patients who showed him the way—not those who died as expected, but those who, against all odds, survived. "I simply began to question why some people didn't die when they were supposed to," he told an interviewer some years later. "I'd ask, 'Why didn't you die?' And I'd hear things like, 'Well, I moved to Colorado. And the mountains were so beautiful, I forgot.' "[74]

Siegel set out to discover what kind of personality it is that "forgets to die." He read through the old psychosomatic literature on cancer-prone personalities, and compared the older cases to his own experiences with patients. His conclusion? There were stable differences between the kinds of patients who survived against the odds and the kinds of patients who in the end did not. The former were openhearted, feisty, and in touch with their needs. The latter were more emotionally repressed—"nice," but not authentic. Siegel also believed, though, that every patient possessed in him- or herself the means to become a survivor if he or she chose to do so—there were "no incurable diseases," he insisted, only "incurable patients."[75] In 1978 Siegel set up the Exceptional Cancer Patient's Program, or ECaP, which aimed to take patients with cancer and help them discover resources inside themselves that might help them to become "exceptional patients." To tap into those resources, ECaP counselors were trained to use a mix of dream analysis, art therapy, and a kind of high-stakes talking therapy called "carefrontation" in which counselors would ask (rather as Weizsäcker in Germany had done in the 1920s): "What benefit are you getting from your cancer? What would you require to give it up?"

Siegel's 1986 book *Love, Medicine, and Miracles,* which eventually sold more than two million copies, pulled everything together, offering a view of cancer and survival that stressed, above all, the need to see cancer as a challenge to patients to heal not just their bodies but their lives:

> Cancer might be called the disease of nice people. They are "nice" by other people's standards, however. They are conditional lovers. They are giving only in order to receive love. If their giving is not rewarded, they are more vulnerable to illness than ever.[76]
>
> Consider some of our common expressions. "He's a pain in the neck/ass. Get off my back. This problem is eating me up alive. You're breaking my heart." The body responds to the mind's messages, whether conscious or unconscious. In general, these may be either "live" or "die" messages.[77]
>
> [W]hen a human being suffers an emotional loss that is not properly dealt with, the body often responds by developing a new growth. It appears that if we can react to loss with personal growth, we can prevent growth gone wrong within us. . . . it is my main job as a doctor to help you develop into a new person so you can resist the unwanted, uncontrolled development of illness.[78]

For many patients, Siegel's vision of cancer was clearly a lifeline. One cancer survivor, a psychotherapist named Janet Collie, put matters this way:

> I am a child of "the Bernie Siegel era." At 30, promptly upon diagnosis, I was handed a magazine interview of this Yale surgeon who said cancer was "God's reset button." I was hooked. I was desperate to save, that is, to heal my total life. It took years for me to sort through the issue of blame vs. responsibility. But I did. I decided that blame is beside the point. So is total responsibility for getting or curing cancer. No one has total control. But responsibility for the fulfillment of one's life, including what one does with one's cancer—that's a different matter. That's a matter of the immune system, maybe even a matter of life and death. . . . You bet I'm grateful to Bernie Siegel for bringing the issue to the forefront of our culture. Because those of us who cut our holistic teeth on that issue developed our own brand of wisdom as a result.[79]

Others, though, found less to admire. In her 1999 book *The Red Devil*, Katherine Russell Rich had this to say:

> After reading *Love*, I suspected that friends and relatives of the stricken, in need of a quick hospital gift, were largely to blame for the sales. I haven't met many actual cancer patients who actually liked the book. The ECaPs, for starters, are obnoxious paragons. "While the typical patient may ask, 'Why me, Lord?' " Siegel writes, "the exceptional patient says . . . 'Try me, Lord.' " It's the reader's patience that's tried as Siegel, who shaves his head bald like a cancer sufferer in what might be construed a PR move, serves up one glowing testimonial after the next—usually to himself. Nurses pop into patients' rooms and demand, "Tell me about Dr. Siegel." People whose tumors mysteriously shrink prior to chemotherapy don't credit their own efforts at, say, positive visualizations, but instead exclaim, "It must be that shiny-headed doctor."[80]

Others were critical for more basic reasons: at least one standard randomized trial had suggested that in fact all these efforts to cure cancer through an honest reckoning of the repressed issues in one's life made no difference. Women with breast cancer who went through Bernie Siegel's ECaP program turned out to have died at the same rate after ten years of follow-up as a group of matched controls.[81]

Nevertheless, the history here is far from over. As recently as 2004, a breast cancer patient in remission told the story—more approvingly than not—about how, when she was first diagnosed in 1993, "my surgeon laughingly told me a story about a patient he'd biopsied for a large lump in her breast. Even before he got the biopsy results back from the lab, he told a colleague he didn't think the lump was malignant, because the woman was 'too mean to have cancer.' "[82] The skeptical conclusions of modern medicine alone seem unlikely anytime soon to spell the demise of the idea of a link between emotional assertiveness and resilience to cancer. The idea persists, and it almost certainly does so in part because, in individual cases, for both doctors and patients, it "feels" true. It persists also, though, for another reason: for many patients it offers a rationale for claiming a different kind of control over the time they still have to live than they would have felt able to claim before—however long in the end that time turns out to be.

Chapter Three

THE POWER OF POSITIVE THINKING

Jesus said to him, If you can believe, all things are possible to him who believes.

—*Mark 9:23*

O my friends there are resources in us on which we have not drawn.

—*Ralph Waldo Emerson, 1838*[1]

On Sunday, December 6, 1998, the *New York Times* published an article about a medical cancer specialist by the name of Dr. Robert Buckman. Buckman had heard about a spectacular cancer treatment while listening to a lecture given by Dr. Bernie Siegel, author of *Love, Medicine, and Miracles*. As the *Times* explained:

> Dr. Siegel told of two oncologists chatting about a study they were participating in to test a combination of four chemotherapy drugs, which had the initials EPHO.
>
> One doctor's patients were doing spectacularly well; three quarters of them were responding to the drugs. But only a quarter of the other doctor's patients were improving. Then the first doctor explained that he had simply rearranged the letters of the drugs so they spelled HOPE.

People in the audience listening to Dr. Siegel were blown away by this. Robert Buckman was one of them. As he put it: "If hope did that, then it was the most powerful anti-cancer agent the world had ever known."

Buckman asked himself, why hadn't he heard of this study before? He wanted to find out more. So he asked Siegel for his source, which turned out to be a book by Norman Cousins (of whom more later). That book in turn took Buckman to an article published in 1988, written by a Californian cancer specialist named William Buchholz. Buckman tracked down his phone number and called him. And Buchholz told Buckman that the story was meant to be a parable; he hadn't intended it to be taken literally. Buchholz said he had simply been trying to tell his colleagues that good clinical practice required more than doling out drugs and avoiding malpractice suits. It was a good moral point to make, but it seemed to have gotten lost in the eagerness to take a different, more literal message from the story. This was a story, it seemed, that was too good not to be true.[2] When I recounted the story of the two oncologists to a large class of Harvard undergraduates and asked how many of them believed it, virtually all the students raised their hands.

Why does a story like this seem plausible to so many of us? The reason is that, consciously or unconsciously, we've been schooled in the assumptions and plotlines of a narrative I call "the power of positive thinking." The message of this narrative is that, no matter how ill a person may be, there's always reason for hope because hope itself can heal. If we can find the strength of mind to stay positive, to believe in the possibility of our recovery—even against all obvious odds—then the strength of our belief can make us well.

Historically, this narrative has deep roots in the Christian tradition. Again and again, the Jesus of the Gospels says to those who seek him out, "Your faith has healed you."[3] Nevertheless, "the power of positive thinking" is not simply a variant of a traditional Christian healing narrative. For I suppose few Christians belonging to mainstream denominations assume that the sick who were healed by Jesus in the Gospel stories got well because they rallied within themselves a certain psychological state—"belief" or "faith"—blessed with intrinsic healing powers. Most forms of Christianity teach that belief opens the door to healing, is perhaps a necessary moral condition for healing; but that healing itself is a gift from God, a reward for faith and a sign of divine power and compassion. In contrast, "the power of positive thinking" narrative puts the focus squarely on the believer and the power he or she possesses in him- or herself. The power that this narrative ultimately asks us to acknowledge is not the power of God but rather the power of faith itself. In this sense, "the power of positive thinking" is a resolutely individualistic miracle narrative. Miracles are possible, it tells us; but it is up to us—up to our capacity to believe—whether or not they happen.

Since the second half of the nineteenth century, there have been two main groups that have invoked stories about the extraordinary power of the human mind to heal itself through faith, hope, and positive thinking. The first of these consists of doctors and scientists who know (or think they know) that all healings, however extraordinary, have a natural explanation. For them, the truth of the power of positive thinking is of great practical interest, but it also functions as a political resource by allowing them to challenge what they take to be outdated or irresponsible supernaturalist understandings of certain kinds of healings. In the modern world, these doctors and scientists insist, there is room for mental but not for frankly supernatural explanations of healing.

The second group of people who have rallied around the power of positive thinking are patients or potential patients who generally do not make such a strict distinction between the powers of the human mind and the powers of God. Instead, this group extols the virtually divine power that exists within each individual, and it actively defies doctors to pronounce on what may or may not be possible. At times, therefore, members of this second group are inclined to offer stories about the healing effects of belief or faith as an explicit challenge to the tenets of conventional medicine and science.

For a long time, these two quite different approaches to stories about "the power of positive thinking" coexisted without a great deal of contact or interaction. It was not until as recently as the 1970s that doctors' professional interests and patients' practical needs finally, if rather uneasily, began to come together. Let us see now how that worked.

Doctors' stories: Lourdes and the "faith cure"

In 1858 a simple peasant girl named Bernadette Soubirous, living in the small village of Lourdes in southwestern France, announced that she had seen an apparition. A "lady" dressed in white had appeared to her and told her she wanted to convey important spiritual messages to the community. First chastised for telling tales, then examined by a medical doctor for signs of delusion or hysteria (none were found), Bernadette was eventually redeemed in the eyes of the authorities when the apparition provided various signs through the child that she was in fact the Virgin Mary, "the Immaculate Conception." The apparition then led the child to a previously unknown spring of fresh water in the back of a grotto; and almost immediately, local people began to report healings after contact with the water. In 1876 the Papacy officially recognized Lourdes as a holy place of healing and pilgrimage.[4]

It did not do so, however, without putting in place some safeguards to ensure that populist enthusiasms for the healing power of Lourdes were kept in check. It would not do for any number of ordinary people to begin deciding for themselves whether or not they were recipients of a miracle healing. Miracles were the sort of things that needed to be determined by experts, and so in the course of recognizing Lourdes as a healing shrine, the Church authorities also undertook to set up a Medical Bureau—a commission of Catholic doctors—whose job it was to act as gatekeepers, to determine which of the many healings at Lourdes really met the stringent criteria for what was called "medical inexplicability." So stringent were the criteria set down that to this day, out of thousands of cases of claimed miracle healings at Lourdes, only sixty-six have been recognized by the Church as true "signs of God." (The most recent was ratified in February 1999: a case of multiple sclerosis suffered by a middle-aged Frenchman named Jean-Pierre Bély.)[5]

Bernadette Soubrious (1844–1879).

© Bettmann / Corbis

From the beginning, the Lourdes Medical Bureau made it clear that

Pilgrims in Lourdes, 1872. The original caption to this image read: "At the Grotto, they drink the miraculous water, admire the miraculous statue and pray for miracles of their own." From A. Deroy, in *Ny Illustrerad Tidning* (November 9, 1872), p. 357, and also *O Novo Mundo* (1873), p. 284. *© Mary Evans Picture Library*

no case brought before it would receive any further consideration if there was even a chance that it was nothing more than a case of hysteria responding to the power of suggestion. In this context, it is worth noticing that the Lourdes Medical Bureau was established in just the same time that Hippolyte Bernheim was developing his new ideas about suggestion, and these ideas were very important to these Catholic doctors as they settled down to their work. They were actually important in two ways: first, they helped to debunk and expose pseudohealings; and second, they called attention to the apparent inability of the new medical understandings of the mind's power over the body to actually account for much of what was going on at Lourdes. Bernheim and others had insisted that suggestion explained hysterical symptoms; but was there any reason to believe that suggestion could explain more? What about the sudden remission of symptoms that were not obviously hysterical in origin? At least one Catholic commentator was clear:

> The most learned and daring of the suggestionists of the present day, Bernheim, a Jew, head of the famous school of Nancy, the more advanced rival of the Ecole de la Salpétrière [*sic*], answers in the negative. . . . Therefore curative suggestion is no explanation [for these cases of remission that do not involve hysteria]. It is not suggestion that operates at Lourdes; the cause which cures acts differently and is infinitely more powerful. . . . As a matter of fact, no natural cause, known or unknown, is sufficient to account for the marvelous cures witnessed at the foot of the celebrated rock where the Virgin Immaculate deigned to appear. They can only be from the intervention of God.[6]

As clerics like these invoked the idea of suggestion only to call attention to its explanatory limits, two former rivals from the world of French hysteria and hypnosis research, Bernheim and Jean-Martin Charcot, found themselves united in mutual indignation. Secularists to the core, with no love for the Catholic Church (Charcot was openly anticlerical, Bernheim was Jewish), both were gravely affronted that members of the clergy should misappropriate medical discussions about the power of suggestion to advance spurious supernaturalist arguments. Neither of these men was prepared to agree that where suggestion let off, God's work might begin. On the contrary, their view of the healings at Lourdes was that these simply showed that medicine had underestimated the power of the mind to heal the body. Alongside the power of suggestion, Bernheim thus began to argue, medicine needed to recognize a second, more potent power of the mind, one that was perhaps often stimulated by religious belief but that in itself had no inherent religious implications. He called this power faith:

> Faith moves mountains, faith performs miracles, because faith is blind, because it does not reason, because it suppresses control and impresses itself directly upon the imagination, without moderating second thoughts.[7]

Charcot took the argument further. Having been finally persuaded of the importance of psychological factors in the genesis of hysteria, he saw no reason to hesitate further simply because Bernheim had been first to see the situation clearly. French medicine needed a strong rebuttal to the Church's views on Lourdes, and he was in a position, on behalf of his profession, to provide one. And so Charcot wrote "The Faith Cure" ("La foi qui guérit"), and arranged for it to be published simultaneously in English and French in 1892, one year before his death.

The article began by setting out its assumptions. The possibility of genuine miracles was not on the table. All the healings at Lourdes, extraordinary as they were, were simply evidence that the natural healing powers of the mind were far more extensive than the medical profession had previously appreciated. Why had it taken Lourdes to reveal this? Charcot focused on the remarkable features of Lourdes as a site. Its remoteness meant that all pilgrims underwent a long, arduous journey to reach it (the train trip from Paris at that time took twenty-two hours). When they finally arrived, they were exhausted and their critical faculties were diminished. Arriving at the grotto itself, they were then immediately immersed in multiple sacred symbols of healing. Joining crowds of other believers, they were infected with the emotional contagion of collective hope. It all added up to a fabulous confluence of factors guaranteed to open the mind to any and all influences. Indeed, others besides Charcot had marveled over this feature of Lourdes: "What medical hypnotist can produce a stage set like this?" exclaimed the physician Félix Regnault.[8]

Citing a case of a patient he himself had seen who had been apparently cured of her tumors by a visit to Lourdes, Charcot argued that the conclusion was clear: either hysteria, known to respond to emotions and suggestion, was a larger category of dysfunction than had previously been thought; or else the mind could extend its influence into the workings of physiology in ways that were still not yet well understood. Either way, medicine needed to get busy, and the Church needed to be put on notice that medicine was on the case. Charcot closed with a little flourish, quoting Shakespeare's *Hamlet*: "there are more things in heaven and earth, Horatio, than are dreamt of in your philosophy."[9]

Bold and straight talking, Charcot's "The Faith Cure" seemed to be just what the doctor had ordered. Some compared the piece to Émile Zola's famous debunking novel *Lourdes*, which had portrayed all the healings at Lourdes as cases of undiagnosed hysteria responding to the manipulations of suggestion and self-deception.[10] Charcot, however, had actually done something more ambiguous and, in a sense, more radical than Zola. No more a believer in the supernatural than Zola, Charcot nonetheless claimed to have been provoked by Lourdes into a conversion of a different sort. Once the embodiment of confident nineteenth-century medical reductionism, he now professed himself awed by the psycho-physiological power of faith, a power capable of producing effects on the body beyond the current horizons of medical understanding. For years afterward, others within the French medical profession, goaded by Charcot, did their best to rise to the challenge of containing and explaining the phenomenon of Lourdes; like Charcot, they were clear that too much was at stake for medical science to simply respond with silence.[11]

Better health through right thinking, part one: Christian Science

In the United States, things unfolded differently. Even though the Catholic minority was keen to benefit from the healing potential of the shrine (by 1900, one hundred thousand bottles of Lourdes holy water were being shipped annually to this country), the American medical profession paid little attention to Lourdes. The few exceptions to this rule can be counted on the fingers of one hand. In 1934, Flanders Dunbar (who would be instrumental in launching psychoanalytic psychosomatic medicine several years later) suggested that the so-called faith cure had a great deal to teach the medical profession, especially about the role played by positive emotions like "confidence" and "contentment" in healing.[12] In 1936, one of America's leading physiologists, Walter B. Cannon, gave a lecture in which he urged his colleagues to pay more attention to the central role played by emotions in disease as a way of combating the threat of what he called "faith healers, Christian Scientists," and "metaphysical Freudians."[13] But there is no evidence that either of these exhortations had much effect on the research and clinical interests of the American medical profession as a whole. Nor, to my knowledge, did any clergymen in the United States make Lourdes the focus of an effort to call people back to the faith. Perhaps Henry Adams had been onto something when he suggested in 1907 that scientifically minded Americans—dazzled more by X rays than cathedrals—did not really respond to the "force of the Virgin."[14]

This is not to say, though, that scientifically minded Americans had no interest in miracle healings; they did. It is just that they took a less institutionalized, more individualistic, and in a sense, more scientific approach to such things. Faith was powerful, but people did not need to wait passively for it to work its miraculous healing effects. Instead, they could study the laws responsible for faith's ability to heal, and then actively apply those laws in their own lives. Miracles could become experiences that people, armed with right understanding, could summon themselves, for themselves. Belief in the possibility of healing could itself produce healing. No Virgin or other intermediary was needed.

Collectively christened the "religion of healthy-mindedness" by William James at the turn of the twentieth century, this way of thinking came more generally to be known as "mind-cure."[15] As a movement, mind-cure drew on the larger currents of antiauthoritarianism and individualism of the time; indeed, it came to function as a kind of early-twentieth-century counterculture.[16] The movement's leaders appealed to a larger popular culture of alternative therapies that collectively aimed to challenge the authority and competence of mainstream medicine.[17] At the same time, they appealed to a larger popular culture of alternative religiosity that was rebelling against the perceived spiritual inadequacies of older forms of Protestantism such as Congregationalism and Episcopalianism.

One striking feature of the mind-cure movement was that both its leadership and its hard-core following included a disproportionate number of women. Given how rare it was for women to occupy prominent roles in society at the time, some have seen the mind-cure movement as an important part of the early history of American feminism. In hindsight, we can understand why it could have come to play such a role. Women were understood to be physically frail and prone to illness, but mind-cure offered them a means of conquering their physical limitations. The minds of women were supposed to be more intuitive than rational, but mind-cure taught that these intuitive qualities were the very ones that opened the door to divine power. In this way, mind-cure empowered women in late-nineteenth-century America to believe that their femininity was—or

Mary Baker Eddy (1821–1910).
Courtesy Schlesinger Library, Radcliffe Institute, Harvard University

could be—a source of strength rather than weakness.[18]

Of the women who rose to leadership in the mind-cure movement in this period, none became more visible than Mary Baker Eddy, the founder of The Church of Christ, Scientist, or Christian Science. Born on a farm in Bow, New Hampshire, in 1821, Eddy was raised in the strict Congregationalist Church of the time but soon began to rebel against some of its less congenial teachings. She was particularly disturbed by the concept of predestination (the idea that the saved and the damned have been predetermined by God), which she found impossible to reconcile with her belief in God as love.

> I was unwilling to be saved, if my brothers and sisters were to be numbered among those who were doomed to perpetual banishment from God. So perturbed was I by the thoughts aroused by this erroneous doctrine, that the family doctor was summoned, and pronounced me stricken with fever.[19]

Over the next several years, Eddy's thinking became increasingly radical. She began to resist any theology that involved an infinitely loving God being willing to allow his creatures to suffer in any way, including from disease. Nevertheless, the fact remained that Eddy's own early adult life was spent in a state of chronic ill health. After the birth of her first child (shortly after the untimely death of her first husband), she became a virtual invalid. When conventional medicine failed to help her, she experimented with various patent medicines and alternative medicines of the day, but to little avail. Then, in 1862, she was told about a clock maker living in Portland, Maine, named Phineas Parkhurst Quimby. Quimby had learned mesmerism from itinerant teachers of the practice and found that a modified version of it—based, he believed, on Christian principles—resulted in remarkable cures.

At first, it seemed that Eddy was destined to become just another of Quimby's success stories. Under his influence, her pain and weakness disappeared and were replaced by a sense of well-being and energy. Within a week, according to her testimony, she was able to climb unaided the 182 steps to the dome of Portland City Hall. She became devoted to Quimby and styled herself one of his students. From him, she learned that he actually did not believe that he truly healed people by manipulating a magnetic fluid inside them. Illness, he said, was caused by people's false beliefs, their failure to realize that the body was a reflection of the mind and that the mind was whole and perfect. These false beliefs themselves were responsible for the symptoms from which people then suffered. Change the mind, correct the beliefs, and the body healed of its own accord. Quimby worked by first creating a mesmeric-style clairvoyant rapport with patients (or so he believed) and then persuading them through the force of his own conviction to change their minds, to begin to think in new, health-inducing ways.

All went well for Eddy until the winter of 1866, when two things happened: Quimby died unexpectedly, and shortly afterward, Eddy slipped on a sheet of ice, fell on her back, and became incapacitated once again. She quickly sought help from one of Quimby's students, Julius Dresser, but he told her he had no time to look after her. Feeling she did not have long to live, she asked for her Bible and read the following passage from the Gospel of Matthew:

> And, behold, they brought to him a man sick of the palsy, lying on a bed: and Jesus seeing their faith said unto the sick of the palsy; Son, be of good cheer; thy sins be forgiven thee. And, behold, certain of the scribes said within themselves, This *man* blasphemeth. And Jesus knowing their thoughts said, Wherefore think ye evil in your hearts? For whether is easier, to say, *Thy* sins be forgiven thee; or to say, Arise, and walk? But that ye may know that the Son of man hath power on earth to forgive sins, (then saith he to the sick of the palsy,)

> Arise, take up thy bed, and go unto thine house. And he arose, and departed to his house.[20]

At that moment, as Eddy later explained, she felt an extraordinary sensation of energy, knew she was healed, and got up from her bed and walked unaided. This experience liberated her, she said, from any lingering attachment to Quimby's approach (whose links to mesmerism she now underscored) and took her back to Christianity proper. Hers was an approach to Christianity that included not only salvation from sin, but from sickness as well. People suffer, Eddy began to teach, simply because they fail to realize the true message of the Gospel healing stories: namely, that the material world, and all the suffering that goes along with it, is an illusion. The fundamental reality of the universe is spiritual; it is divine Mind or God, in which all of us participate and in which we have health and wholeness. Arise and take up your bed—because whether you know it or not, you are already well.

The book *Science and Health*, written and rewritten over many years, was Eddy's attempt to explain all of this in detail. First published in 1875, it became and remains the foundational text of Christian Science, the "key" to the Bible, read side by side with it in Christian Science congregations every week. Eddy's church, the Church of Christ, Scientist, was established in 1879. And her Massachusetts Metaphysical College, dedicated to training people to work as Christian Science "practitioners," or healers, received its charter from the state in 1881.[21] It was not supposed that practitioners healed others directly; their job was to help their patients overcome the illusions that cause them to suffer and, through prayer and dialogue, enable them to experience their true natures, not as material creatures but as ideas of divine Mind.

Christian Science reached its peak of influence as a church in the 1930s, when its membership was estimated at close to 250,000.[22] During these years, its public image was of a rather secretive organization that allegedly encouraged parents to deny their children essential medical care.[23] More recently, as its membership has declined, it has attempted to become more relevant to modern concerns; and under the leadership of Chairman Virginia Harris, it worked in particular to develop a public image as a radical voice for the power of mind-body healing. Mary Baker

Eddy herself, Virginia Harris proposed, was a "pioneer in the recognition of the mind/body connections" long before the rest of medicine was willing to recognize their existence.[24]

Christian Science practitioners today dialogue freely with the medical profession and are happy to acknowledge a common interest in mind-body interactions, but ultimately insist their own understandings take them beyond anything that secular medicine would ever countenance. They emphasize that these understandings are not just rooted in blind faith, but grounded in repeated personal experience. It is in this sense above all, they say, that Christian Science should be seen as a science, a form of Christianity that aims to demonstrate its truths through empirical demonstration. A practitioner, speaking to a class of Harvard undergraduates, put the matter as follows:

> When my sister was in an auto accident, she was told that if she didn't have an immediate operation to remove the glass from her head she would die. It wasn't a difficult decision for her to decide to come home instead of going to a hospital, because it was natural for her to trust the power of prayer. And it was effective. That very night, the glass came out of her head, just naturally, the wounds healed up, she was back in school in a few days.
>
> When I see cures that defy any known laws of physics or medicine, and I've seen many, it's obvious that matter isn't the solid substance it appears to be. Something is going on that isn't explained by the laws of physics or modern medicine. So our laws of physical science, which we're so proud of in our modern civilization, have overlooked something very significant. I can't just shrug that off; I want to know what's missing that is so powerful it can restore health to people considered medically hopeless.[25]

Better health through right thinking, part two: New Thought

From the beginning, Christian Science had a rival, or rather a cluster of rivals, with roots in the same late-nineteenth-century mix of religious and medical discontents as Christian Science: Unity Church, Religious Science, Divine Science, Mind Science—all of these were different semi-institutionalized variants of mind-cure that, by 1915 (when leaders gathered for a conference), had loosely united under the umbrella term "New Thought." All of the different churches of New Thought saw themselves as Christian in spirit and essential doctrinal orientation; many of them were established by dissident onetime followers of Mary Baker Eddy. All recognized Phineas Parkhurst Quimby as their founding father, and all insisted that Christian Science had failed properly to acknowledge its own equally deep debt to Quimby's mind-cure system. Some went so far as to accuse Eddy of taking ideas straight out of Quimby's manuscripts and putting them into *Science and Health*.[26] Beyond this, the leaders of New Thought drew variously and generously on a great many esoteric, pantheistic, and occult traditions of the time, including Swedenborgianism, Rosicrucianism, Theosophy, Anthroposophy, and Transcendentalism.[27]

Sometimes the resulting mix could produce styles of exposition that were anything but clear. But behind the overblown and occasionally febrile rhetoric was a quite simple two-part message: first, people possess great powers deep within themselves on which they can draw for health and healing; and second, they can open the door to those powers by filling their minds completely with the conviction of wellness, and refusing to dwell on anything negative. In the words of Ralph Waldo Trine, a popular New Thought writer of the time:

> Don't talk of sickness and disease. By talking of these you do yourself harm and you do harm to those who listen to you. . . . Never affirm or repeat about your health what you do not wish to be true. . . . Stoutly affirm your superiority over bodily ills, and do not acknowledge yourself the slave of any inferior power.[28]

Emma Curtis Hopkins, a onetime follower of Christian Science who went on to become one of the most prominent female New Thought teachers (she taught many who became teachers in their own right), summed up the message of New Thought this way:

> All nature will say "Amen!", if you proclaim that the Good you are seeking is free health. . . . Sometimes when you say to the sick man,

> mentally, that the Good he is seeking is his God, and God is free health, he will get well in five minutes. His mind was unconsciously groping around for the Divine Word that could heal him, and you spoke for him.[29]

How should one claim one's "free health"? Most New Thought texts offered specific visualizations, mantras, and prayers to help people open their minds to the power within them. Most of these techniques were generic, but a few were quite specific, as evidenced by this charming "Prayer for the Dyspeptic," published in the 1887 *Mind-Cure Mentor*.

> Holy Reality. We BELIEVE in Thee that Thou are EVERYWHERE present. We *really* believe it. Blessed reality we do not pretend to belief. WE BELIEVE. Believing that Thou are everywhere present, we believe that Thou are in the patient's stomach. Help us to stoutly affirm with our hand in Your hand, with our eyes fixed on Thee, that we have no Dyspepsia, that we never had Dyspepsia, that we will never have Dyspepsia, and that there is no such thing, that there will never be any such thing. Amen.[30]

Did all the varied exercises designed to cultivate wellness deliver the goods? New Thought certainly counted many happy customers, and William James, the Harvard psychologist who was a serious student of the movement, had no doubt that it deserved to be taken seriously: "The greatest discovery of my generation is that man can alter his life simply by altering his attitude of mind," he opined. The mind-cure movements, more clearly than anything else, had proven this:

> The blind have been made to see, the halt to walk; lifelong invalids have had their health restored. The moral fruits have been no less remarkable. The deliberate adoption of a healthy-minded attitude has proved possible to many who never supposed they had it in them; regeneration of character has gone on, on an extensive scale; and cheerfulness has been restored to countless homes.[31]

If New Thought could have such apparently potent effects on health and general well-being, could it do more besides? Could it, for example, make a person not just healthy but also wealthy? By the early twentieth century, an increasing number of New Thought teachers were saying it could. Wealth, no less than health and happiness, was one's divine birthright, and New Thought's mental technology could be used to get one's share. Wallace Wattles's classic text *The Science of Getting Rich*, first published in 1910, spelled out the position in the form of a few pungent aphorisms:

> It is the desire of God that you should get rich. He wants you to get rich because he can express himself better through you if you have plenty of things to use in giving him expression. He can live more in you if you have unlimited command of the means of life.
>
> The universe desires you to have everything you want to have.
>
> Nature is friendly to your plans.
>
> Everything is naturally for you.
>
> Make up your mind that this is true.[32]

Automobile magnate Henry Ford was so persuaded of the power of New Thought to facilitate worldly success that he ordered bulk copies of Ralph Waldo Trine's *In Tune with the Infinite* and had them distributed to various high-profile industrialists. His famous comment "If you think you can, you can. And if you think you can't, you're right," is New Thought thinking tailored to the no-nonsense world of young capitalist America.

During the Depression years, the new prosperity-oriented New Thought was more popular than ever. Dale Carnegie's *How to Win Friends and Influence People*, Napoleon Hill's *Think and Grow Rich*, and many others became must-reads for company executives across the country. Another publishing sensation of the Depression years was the English translation of Émile Coué's *Self Mastery through Conscious Autosuggestion*, the book that taught people to repeat, twenty times a day, the still well known mantra "Every day in every way I am getting better and better."[33]

Even children in these days were offered age-appropriate versions of the new success-minded New Thought. In 1906, a Sunday school publication called *Wellsprings for Young People* published a little story called

"Thinking One Can." The story was about a little locomotive that agreed to pull a heavy load over a great hill after all the big engines refused to try—and which succeeded because it believed it would. In 1930, the publishing house Platt & Munk released a version of the story (which had been retold several times since 1906) under the title *The Little Engine That Could*. In this version, the little engine struggles up the hill chanting its mantra, "I think I can, I think I can," over and over, until it succeeds. The book remains a best-selling title to this day.[34]

From New Thought to the power of positive thinking

After World War II, the United States experienced a new prosperity boom, and New Thought found a new champion, Norman Vincent Peale. With his mainstream credentials and plain-talking style, Peale did more than anyone else to dissociate the basic New Thought message from its original, somewhat esoteric roots and make it feel as American as baseball and apple pie. Pastor for fifty-two years of the Marble Collegiate Church in New York City, Peale made his reputation in 1952 with publication of the runaway best-seller *The Power of Positive Thinking*, which opened with the ringing words "Believe in yourself! Have faith in your abilities!"[35]

Norman Vincent Peale (1898–1993), photographed August 23, 1946.

What made Peale such an effective spokesman for positive thinking in the postwar era? Peale had trained as a Methodist preacher but imbued with the larger "therapeutic" impulse that was beginning to dominate in American churches in the postwar era, he had grown dissatisfied with the fundamentalist Christian doctrines he had learned in seminary.[36] In 1937 he had conspired with a Freudian psychiatrist, Dr. Smiley Blanton, to set up a psychotherapy clinic in the basement of the Marble

Collegiate Church, where Peale was the pastor.[37] At the same time, Peale began reading the older New Thought and Christian Science literature for new ideas. In fact, the term "the power of positive thinking" was not his own coinage, but was taken from a New Thought text by one Charles Fillmore.[38]

In the end, though, Peale did find his own uniquely effective formula. Like conventional New Thought writers, Peale believed a person could use the power of positive thinking for anything—from renewing health to getting a better job to saving one's marriage. Unlike those older writers, however, he self-consciously connected his message to the modern therapeutic ideas of Freud and the latest thinking in psychosomatic medicine. Sin, Peale wrote in one of his later books, is "a splinter in the unconscious mind" that can fester and cause ill health if not attended to. "Emotional stress," he explained, plays a role in all illnesses, "from the common cold to cancer."[39] Once these truths were recognized, he concluded again and again, all such problems could be countered with positive thinking, the great secret of psychology and the Gospels alike:

> [S]ome people who firmly believe that Jesus Christ healed the sick in the First Century find it difficult to believe that this same power operates today and especially for them. The age of miracles is past, they say sadly. Healing is now done through scientific medical means . . . (and usually they add piously) though faith helps. . . . But Divine healing is "scientific" in that it is conditioned by law, spiritual law, the highest form of law.[40]

The proof of the power, he concluded, was everywhere to be seen:

> Smith has never again had need to revert to the habit of taking tablets. He learned the amazing power of positive thinking to heal. Let me repeat. The technique is to believe that you are going to be better, believe that positive thinking is going to work for you, and remedial forces actually will be set in motion.[41]

Since its publication, *The Power of Positive Thinking* has sold some twenty million copies, and it continues to sell three thousand copies weekly. Shortly after its publication, it was number two on the best-seller list for two years in a row; the number-one best-selling book from that period was the Revised Standard Version of the Bible![42] In addition, at the height of his popularity Peale was reaching millions of people through his weekly radio program, "The Art of Living," which was broadcast on the National Broadcasting Company (NBC) for fifty-four years. His sermons were mailed to 750,000 people a month, and 4,000 people attended his two Sunday sermons at Marble Church every week. His *Guideposts* magazine had a circulation of more than 4.5 million.[43]

At the same time, Pealism—as some called it—was obviously far from an orthodox way of understanding the Christian message. Peale faced a storm of criticism from leaders of other Protestant congregations, especially in the early years of his ministry. It was one thing when institutions on the margins of mainstream Christianity, like Christian Science, preached the power of right thinking to heal. It was another thing altogether when a minister of a highly visible mainstream parish preached such a message. "Peale speaks much of faith," an author writing in 1957 for *Christianity Today* fumed, "but it is not faith in God, but 'faith in faith,' which means in your capacities." The author went on:

> This is neither religion, moralism, or anything more than self-help baptized with a sprinkling of devout-plus-medical phrases. For those who believe in the God of Scripture, the reality of vitality, of good and evil, and the grace of God unto salvation, there is nothing here but the frenzy of a guilty life and the misery of creeping death.[44]

Such criticisms hurt, and at one point Peale felt so bruised he considered resigning his pastorate. Only the support of his family, particularly his father, emboldened him to continue. Over time, the voices of critics like these were muted against a backdrop of general public acclaim for his message. Peale's interpretation of Christianity may indeed have turned sin into a "splinter in the unconscious mind," and might not have been too troubled about repentance or salvation. It seems, however, to have been just what large numbers of Americans were hungry to hear. By the 1960s, Peale had taken his place in American mainstream society as a beloved and trusted religious populist—he was even chosen to conduct the marriage ceremony between U.S. president Richard Nixon's daughter Julie and David Eisenhower. In 1966, America's other great preacher, Billy Graham, stood up before representatives from the National Council of Churches and told them, "I don't know anyone who has done more for the kingdom of God than Norman and Ruth Peale."[45] It is hard to imagine a more striking acknowledgment of the mainstreaming of positive thinking, at least in American religious culture.

The medicalization of positive thinking

Peale brought the gospel of positive thinking into the mainstream of American religious culture. But he did not succeed in bringing it into the mainstream of American medicine and science. In spite of positive thinking's overwhelming emphasis on physical healing, mainstream American medicine barely paid any attention to the claims and testimonies of its teachers and followers. Even American psychosomatic medicine, focused as it was on Freud and psychoanalytic perspectives, seems barely to have registered the existence of the faith healers and positive thinkers. Indeed, the only reference to positive thinking to appear in the journal *Psychosomatic Medicine* before 1970 was in an article from 1962 by George Vaillant on psychosomatic aspects of schizophrenia. Here Vaillant made reference to the importance of "confidence and faith" in certain therapeutic processes, only to note wryly that he was aware that such encouragements would sound "pretty banal" to readers of that journal. After all, he said, "we are physicians, not purveyors of positive thinking."[46]

The general sense that positive thinking was a fine ideology for self-help gurus and preachers but of no interest to serious medicine did not begin to change until the second half of the 1970s. Why then? In 1976, the editor of the *Saturday Review* and well-known liberal political analyst Norman Cousins authored a highly unusual paper that was published in the prestigious *New England Journal of Medicine*. The paper was called "Anatomy of an Illness (as Perceived by the Patient)," and it was an autobiographical account of a radical experiment in self-healing that Cousins had undertaken in the 1960s. Diagnosed by his doctors with a degenerative disorder called ankylosing spondylitis (which causes the breakdown of collagen, the fibrous tissue that binds together the body's cells) and given a grim prognosis (a 1 in 500 chance of recovery), Cousins had decided there was nothing to be lost and potentially much to be gained by jettisoning the procedures of conventional medicine and taking control of his own treatment.

Significantly, however, Cousins claimed to have taken as his starting point for his alternative treatment, not the works of Mary Baker Eddy or Norman Vincent Peale, but the laboratory research of scientists such as endocrinologist Hans Selye and physiologist Walter B. Cannon on "the negative effects of the negative emotions on body chemistry." Reviewing the findings of these men, Cousins put it to the readers of *NEJM* that they might have told only half the story:

> What about the positive emotions? If negative emotions produce negative chemical changes in the body, wouldn't the positive emotions produce positive chemical changes? Is it possible that love, hope, faith, laughter, confidence and the will to live have therapeutic value? Do chemical changes occur only on the downside?[47]

Cousins decided it was in his interest to find out. With the support of an open-minded doctor, he arranged to be checked out of the hospital and into a hotel ("I had a fast-growing conviction that a hospital was no place for a person who was seriously ill"), where he began to manage his own recovery. The sense of control this produced in its own right was, he believed, health inducing. "Since I didn't accept the verdict [of no recovery], I wasn't trapped in the cycle of fear, depression and panic that frequently accompanies a supposedly incurable illness."[48] He took himself off all pain medications and put himself instead on a diet of steady positive effect: Marx Brothers films, reruns from the television show *Candid Camera*, and humorous literature.

The first evidence that the "laughter" treatment was working, he reported, came when he discovered that "10 minutes of genuine belly laughter had an anesthetic effect and would give me at least two hours of pain-free sleep." This was just the beginning. Over a period of mere weeks, a combination of "the laughter routine" with massive injections of vitamin C led to a rapid remission of virtually all of his debilitating symptoms. The account he offered of his remarkable recovery ended with a description of himself—who had once been nearly paralyzed—standing in the surf in Puerto Rico, then jogging on the beach, and then finally back at work full-time at the *Saturday Review*. This was a story as good as any to be found in countless New Thought texts since the turn of the century, but what gave it particular authority was that it was published in the most prestigious mainstream medical journal of its day.[49] And it certainly did not hurt that the man who was telling the story was a prominent public intellectual. "If Joe Blow of Altoona" had sent in an article of this sort to the *NEJM*, fumed one of Cousins's rare critics, "it wouldn't even get the courtesy of a rejection."[50]

That might well have been true, but it is probably also true that even a Norman Cousins might have had trouble winning a hearing in the *NEJM* before the 1970s. This is because in the 1970s, mainstream medicine felt itself under greater pressure from the general public than at any time since the late nineteenth century. There was a general sense that the medical profession now cared more about serving its own voracious professional ambitions than it did about serving the legitimate health needs of patients; and there was much discussion of the idea that in many cases medicine actually did more harm than good.[51] As discontent intensified, the appeal of alternatives outside the mainstream also grew. These years saw an exponential growth of interest in meditation, acupuncture (recently introduced from China as one legacy of then-President Nixon's diplomatic efforts), jogging, herbal treatments, biofeedback, and more. Such treatments were widely seen as gentle, effective therapies, but also as sources of patient empowerment that could ultimately have a transforming effect on the doctor-patient relationship and hence on medicine as a whole. "Biofeedback promises to return us to a more holistic kind of medicine," proclaimed one typical article from the period, "in which the patient will acquire more responsibility for, and power over, his own health. . . . Biofeedback puts the emphasis back on training, rather than the 'miracle pill' or surgery."[52]

Against this backdrop, Cousins's sudden arrival on the scene looked like a lifeline to many in the medical profession. Here was a patient interested in exploring holistic alternatives to the mainstream, but who told the mainstream medical profession that he wanted to ally with rather than fight them. He made a point of praising his own physician as an open-minded man who was willing to work with him, and implied that partnerships of the sort he had experienced might be ideal for many patients. Most important, perhaps, he offered a now-receptive medical profession an opportunity to do the same kind of thing that Charcot, eighty years earlier, had proposed should happen: to claim for itself the world of religious healing and faith that otherwise kept tempting patients away from sensible medical care; and to do so by subjecting the claims made by faith to the sober light of scientific investigation.

In a later book-length version of his story, also called *Anatomy of an Illness as Perceived by the Patient*, Cousins reported he had received some three thousand letters from physicians in the wake of his original article, virtually all of which praised his courage and the approach he had taken in orchestrating his unorthodox treatment.[53] Within a few years, Cousins had accepted an invitation to join the medical faculty of the University of California in Los Angeles as Adjunct Professor of Medical Humanities, where he proceeded to oversee research on the biochemistry of healing and the emotions. In the 1980s, he spearheaded a task force to explore the medical potential of the emerging field of psychoneuroimmunology. The Norman Cousins Center for Psychoneuroimmunology is still active at UCLA today.

The positive power of the placebo effect, part one

Something else changed when the power of positive thinking was claimed by mainstream medicine. Placebos, previously a key prop in skeptical stories about the power of suggestion, suddenly had the potential to become something new. At one point in his article, Cousins noted that some skeptics of his story had suggested that everything that had happened to him had been a "mere" placebo effect. Well, Cousins said, if that were true, then did it not imply that it was time to take a fresh look at placebo effects? Given the magnitude of the changes in health that he had personally experienced, on what possible basis could medicine assume that effects of placebos were either ephemeral or insubstantial? A bit like Charcot eighty years earlier, Cousins invoked the object lesson of Lourdes. But he did so in a way that suggested that what was once strictly a doctors' insight into the mechanisms behind the faith cure could now be shared by patients:

> It is quite possible that . . . everything I did . . . was a demonstration of the placebo effect. If so, it would be . . . important to probe into the nature of this psychosomatic phenomenon. At this point, of course, we are opening a very wide door, perhaps even a Pandora's box. The vaunted "miracle cures" that abound in the literature of all the great religions, or the speculations of Charcot and Freud about conversion hysteria, or the Lourdes phenomena—all say something about the ability of the patient, properly motivated or stimulated, to participate actively in extraordinary reversals of disease and disabilities.[54]

Was the placebo effect, then, the faith cure of our time and indeed the key to making sense of all faith cures, past and present? In the mid-1970s, two events unfolding within a few years of each other seemed to suggest that the answer to this question could be yes. The first was an outgrowth of the discovery in the early 1970s of endorphins, substances in the brain that are chemically similar to opioids and that function as the brain's own natural "painkillers." In 1978 a report was published which suggested that placebo treatments for pain were mediated, at least in part, by these biochemical substances. When an opioid blocker called naloxone was used to block the body's opioid receptors (without telling patients this was being done), placebo responders stopped reporting relief of pain.[55] Was there then a real biochemistry to placebo responses—and, by extension, to the power of positive thinking—just as Cousins had predicted?

The second event that began to transform the placebo from a suspect instrument of suggestion to a catalyst for the miracle cures of positive thinking was the rise of the new field of psychoneuroimmunology (see chapter four for more). In 1975, psychologist Robert Ader at the University of Rochester put a powerful immune-suppressing drug, cyclophosphamide, in saccharine water and fed it to rats. His goal was to create a state of nausea in the rats and condition them to associate it with the sweet taste of the water; he had not originally realized that the drug also suppressed the immune system. However, when his rats grew ill and began to die, he realized something was wrong. By this time, he had stopped offering them the tainted water and was just feeding them plain saccharine water. Nevertheless, the rats continued to grow ill and die.[56]

Why was this happening? In the end, Ader, having informed himself about the biochemical effects of cyclophosphamide, recruited an immunologist named Nicholas Cohen to help him find out. The two men concluded that the rats had indeed been conditioned. The conditioning had happened, however, in a way that then-current understandings in physiology had said was impossible. The rats' immune systems had been conditioned; that is, they had continued to act as if they were still being suppressed by the drug. Faced with his unexpected results, Ader and Cohen began to ask whether his saccharine solution was nothing more or less than a placebo version of cyclophosphamide. And if it was, wasn't it also possible that at least some placebo effects in humans were similarly produced and might result in bodily changes no less potent than the effect he had stumbled on in the animal laboratory?

Intermezzo: Cancer, AIDS, and the politics of positive thinking

It was an exciting thought. However, this new laboratory work on the placebo effect did not immediately change the focus and tone of discussions about the power of positive thinking. The reasons for this are not hard to understand. In the 1970s, positive thinking was part of the new, patient-empowering holistic medicine. Placebos, whether handed out in a clinic to pacify annoying patients or used in trials to test new drugs, were identified by critics with the most patronizing and ethically fraught face of mainstream medicine.[57] A lot would have to change before something that was once seen as disempowering patients could be perceived as the key to treatments that might actually empower them.

In the meantime, the medicalization of positive-thinking ideas proceeded on other fronts. In the 1980s, many disenchanted cancer patients took heart from the new ideas coming out of psychoneuroimmunology and embraced technologies involving positive thinking as a possible way of boosting the ability of their own immune systems to fight their cancer. Visual imagery exercises in which patients imagined their desired outcome—a technique widely advocated by the old New Thought teachers—became a particularly prominent part of this development. *Getting Well Again*, a best-selling book first published 1978, became the manual of choice for patients attracted to this tactic.[58] Here, patients learned to imagine the cancer cells inside their bodies as (among other things) weak and confused entities that were being systematically destroyed by an army of courageous white blood cells sent out to fight the good fight by their immune systems.

These developments interacted in the thinking of both patients and psychologically minded doctors with new quasi-psychoanalytic personality research on cancer that partially redefined what it might mean to speak of positive thinking. In the 1970s, researchers had followed a large cohort of breast cancer patients over a period of ten years to see if there was a relationship between different kinds of "coping styles" and disease outcome. The results indicated that patients who refused to believe their disease could not be beaten—who exhibited what people began to call a "fighting spirit"—had better outcomes than women who had a fatalistic attitude toward their diagnosis.[59]

Testimonials to the power of a positive fighting spirit, coupled with diligent visualizations, began now to circulate in the self-help literature, patient memoirs, and on internet sites:

> Lisa has the office across the hall from mine, where she is an attorney for Legal Services for Prisoners. When I first knew her, she did not tell anyone that she had leukemia. As she puts it, other people's reactions are often less than helpful. However, she had the Simontons' book, *Getting Well Again*, and was doing visualization and following their guidelines on her own.
>
> . . . Lisa began working on the tumor with visualization, and she and I also did a visualization session. Once she began visualizing her white cells attacking it and eliminating it totally, as a blueprint signaling her intentions to her body, it disappeared in a very short time. It remains absent to this date, more than a year later. Interestingly,

in spite of the biopsy and all the X-rays, her physicians now deny that this tumor was ever there. It simply does not fit their belief system of what can happen.

Lisa is in remission again and happily married. Her energy, her strong positivity in every aspect of her life, and her unique blend of compassion and humor unleash a life force and energy that sustain her through everything.[60]

Cancer, however, was not the only place where positive thinking became a prized psychological tool—and expression of rebellion—for patients who felt let down by their doctors and were looking for a medically sound means of self-cure. In the second half of the 1980s, as the AIDS epidemic hit and mainstream medicine failed to offer patients effective treatments, positive thinking emerged here too as an important part of the grassroots response to the disease, particularly within the hard-hit American gay community. Believing they had been abandoned by the mainstream medical profession (in part, many thought, because people did not really care if homosexuals died), a radicalized community of HIV-positive patients decided they would simply refuse to believe they were going to die. If the new research was true that the brain and the immune system were in constant communication, they of all people—with grossly compromised immune systems—needed to be sending only messages of survival to their bodies. Think negative thoughts, some patients said, accept the death sentence pinned onto you by the medical community, and you effectively condemn yourself to a self-fulfilling prophecy. In 1988, one activist AIDS patient in New York City put the matter plainly in the pages of *New York Native*, a gay and lesbian periodical:

> Psychological warfare is being waged against gay men in the United States. For the past month or so the media have been disseminating hostile propaganda, with the message that we will all die, that we must die. These death threats do not issue from the usual bigots. . . . We are being cursed in the name of science, and the imprecations directed against us have the imprimatur of the Public Health Service (PHS). The prognosis of doom is emanating from that peculiar form of medical survey research known as "epidemiology."[61]

That same year, 1988, psychiatrist Sanford Cohen offered intimate clinical evidence that the AIDS patient was in fact vulnerable to being killed by negative beliefs. In an article about AIDS and "voodoo death," he described a patient whose mother

> learned on the same day that her son was gay and had AIDS. She reacted to this with hostility and openly maintained a prayer vigil outside the intensive care unit, praying that her son would die because of the shame he had caused her. The patient could hear his mother praying. One hour later the patient died, much to the surprise of his physician, since he did not appear to be terminal.[62]

If both doctors and patients agreed that negative beliefs and the death wishes of others could literally kill, then the response seemed obvious too: refuse, simply refuse, to grant such beliefs and wishes any potency; fight them with defiant positivity. As early as 1986, two psychologists involved in promoting frontline psychosocial responses to the AIDS crisis, Lydia Temoshok and George Solomon, described a patient who, in spite of significant T-cell degeneration, was remarkably healthy. Why? They couldn't say for sure, but they had been struck, they said, by the patient's "superb attitude, determination, 'fighting spirit,' . . . and other psychosocial attributes." It seemed reasonable to suppose that these striking qualities had played "a significant role" in stabilizing this man's condition.[63]

For some, the call to positive thinking became the basis for a grassroots political movement of its own. In 1994, a Pennsylvania newspaper profiled a new patient-advocacy AIDS group that called itself "Positively HIV." Its founder, an HIV-positive gay man named Gary Hite, was described by the journalist as someone who

> defies the stereotype of a person who is HIV positive. With his large build, he's robust, not emaciated. With his sharp mind and engaging personality, he's energetic, not sluggish. . . . Hite also relies on his upbeat attitude, "I accept the fact that I'm HIV positive. Others can, too. If people want to die from AIDS, I won't stop them. But we suggest that people who are HIV positive can choose to live."[64]

The rallying call to positivity was inspiring, perhaps, but how did it sit with those patients who failed to maintain the robust optimism of a Gary Hite or who, unlike Lisa, did not seem able to cause their tumors to melt away under the laser-beam force of their upbeat visualizations? An article published in the *New York Times* in 1988—a so-called report from the front lines of the cancer wars—gives us some sense:

> I was a positive thinker, blessed with discipline and dignity. I meditated, visualized and prayed. No fool I, I covered all bases, leaving nothing to chance. I believed in the mind-body connection, and believing in it as I did, I was fairly sure I'd be safe.
>
> Wrong. Wrong. Wrong. Gradually, imperceptibly at the beginning, my condition deteriorated. Pains intensified, and walking, which had been difficult under the best of circumstances, became, within a brief time, virtually impossible.

What did the patient do, confronted with this failure?

> All right, I announced grudgingly. My way didn't work. I'll try theirs instead. And so I did. For months, I was bombarded by treatments, medical, scientific treatments that assaulted my body, but that touched my soul not one bit. It wasn't long before I discovered that I . . . longed for the spiritual component that was now gone. . . . I was more than a body. If I didn't address the other issues of my life . . . I'd have no reason to get better.

In the end, this particular patient found a middle ground for herself, concluding: "I wasn't wrong three years ago, just naive. Nothing is as important as believing in oneself, but one has to keep right on believing when the going gets rougher than you ever imagined it could. . . . My mistake, in fact," she concluded, "had been to read about survivors who'd made it look so easy. It always sounded as though they'd been a breath away from death one minute and happily swimming laps the next."[65]

Clearly it wasn't so easy—and just maybe the whole positive thinking approach did more harm than good. In 1985, Marcia Angell, one of the editors of the *New England Journal of Medicine*—the same journal that had published Cousins's article a decade earlier—wrote an editorial in which she lashed out against a self-help publishing industry that promised every desperately ill person that he or she, just by changing his or her attitude, could beat the odds:

> . . . we are told that just as mental state causes diseases, so can changes in our outlook and approach to life restore health. . . . Is there any harm in this belief, apart from its lack of scientific substantiation? It might be argued that it allows patients some sense of control over their disease. If, for example, patients believe that imagery can help arrest cancer, then they feel less helpless; there is something they can do.
>
> On the other hand, if cancer spreads, despite every attempt to think positively, is the patient at fault? It might seem so. . . . After all, a view that attaches credit to patients for controlling their disease also implies blame for the progression of the disease. . . . In addition to the anguish of personal failure, a further harm to such patients is that they may come to see medical care as largely irrelevant . . . and give themselves over completely to some method of thought control.[66]

In 1985, when Angell was writing, she probably spoke less for a community of disappointed patients and more for a community of mainstream oncologists and other cancer workers, many of whom now felt besieged and undermined by the strength of the popular trend toward mind-body thinking about cancer. There is no evidence that her slap at this trend within the pages of a specialist journal made any significant dent in the sales of the relevant paperback best-sellers.

There is other evidence, however, that by the late 1990s initial popular enthusiasms for dissolving tumors through positive attitudes had begun to temper. In 1999, a great deal of publicity was given to an article published in *The Lancet* that found no significant effect of "fighting spirit" on five-year survival rates from cancer; less attention was paid to the fact that this study also found "a significantly increased risk of relapse or death" in women who scored high on measures of "helplessness and hopelessness" and depression.[67] A poll of 957 Americans commissioned by the American Cancer Society in 2005 found that, while many "false beliefs" about cancer still abounded, 89 percent of those polled disagreed with the statement that "all you need to beat cancer is a positive attitude."[68] It is possible that many who said this would still have agreed that positive attitude was one of several important factors in beating cancer, but the poll did not allow for that level of nuance.

The positive power of the placebo effect, part two

What the power of positive thinking may have begun to lose by the late 1990s in terms of political stridency, it gained in terms of biomedical respectability. More than anything else, this respectability was gained through the increasingly firm identification of positive thinking with the placebo effect. Between 1997 and 2000, there appeared no fewer than five academic books on the placebo effect (one of which I edited), all of which took up the Norman Cousins challenge from 1976 in various ways. The main argument offered by this new literature was that the placebo effect was important above all for what it taught us about self-healing. It was not just a trick; it produced real (physiologically discernible) effects, and therefore it need not inherently undermine patient trust in doctors or function as a source of patient disempowerment. On the contrary, rightly understood, it could be a source of patient empowerment and a means of enriching the doctor-patient relationship.[69]

The popular media were fascinated. The cover story in the January 9, 2000 issue of the *New York Times Magazine* set the tone. Titled "The Placebo Prescription," it enticed readers with the titillating banner "Astonishing Medical Fact: Placebos Work! So Why Not Use Them as Medicine?"[70] The article revisited the classic research on the physiology and mechanisms of placebo action from the 1970s. It polled a range of experts about the clinical scope of the effect, how it might be used, and what the most important directions for new scientific research might be. But its real showstopper was a true-life story that showed the "astonishing" placebo effect in action.

Here is how the story went. In the early 1990s, a surgeon by the name of Bruce Moseley persuaded his hospital to allow him to conduct a placebo-controlled trial of a common form of arthroscopic knee surgery. Surgical techniques are rarely subjected to controlled trials since the inherent risks of surgery are considered too great to justify asking patients to submit to them without clear benefit. Nevertheless, in this instance Moseley prevailed. Patients were recruited for a trial and then divided into two groups. One group underwent the usual surgical procedure. The other group was subjected to all the paraphernalia and theater of a real operation, but Moseley actually did not cut, scrape, or do anything therapeutic to their knees at all—he just opened them up and then closed them again. He himself did not know whether or not he would be "really" operating until he opened an envelope in the operating theater. What was astonishing about all this was that the patients who received fake treatment improved: not just briefly or subjectively, but on multiple measures of objective functioning. Even after they were told they had received the placebo version of the surgery, they continued to walk better, declared they slept more soundly, reported they were able to mow the lawn again, and more. One seventy-six-year-old World War II veteran who underwent the placebo version of the surgery summed up his satisfaction this way:

> I was very impressed with him [the surgeon], especially when I heard he was the team doctor with the Rockets [a U.S. basketball team]. So, sure, I went ahead and signed up for this new thing he was doing. . . . The surgery was two years ago and the knee never has bothered me since. It's just like my other knee now. I give a whole lot of credit to Dr. Moseley. Whenever I see him on the TV during a basketball game, I call the wife in and say, "Hey, there's the doctor that fixed my knee!"[71]

The *New York Times* article generated a lot of copycat journalism, and a huge amount of interest across the country. Again and again, the message went out: It's real! It works! How can we use it? Having become involved in this area myself, I was often asked for my opinions on all the excitement, but never more poignantly than in the winter of 2000, when I received the following email from a young man in South America (I have deleted the author's name to protect his privacy):

> My name is . . . , I live in Cali, Colombia. A few days ago I listened [to a talk] about the placebo effect. My father has cancer, and I believe a placebo can help him.
>
> I need to know where I can get the sugars pills or saline injections, and how I must to give them to him.
>
> My father's doctors have said that his cancer is very bad, and it's so difficult to treat. He has been following a treatment for 8 years, but he always falls ill again. I think the placebo effect is his last hope.
>
> If you can help me, I'll be grateful all my life.

How did we arrive here? How could something once so suspect, so reviled, now emerge as the "last hope" of a loving son trying to save his father? What are we to make of the fact that, in 2001, we would even see a self-help publication about the placebo effect? *Faith and the Placebo Effect* did not mince words: "The placebo effect is the good news of our time. It says, 'You have been cured by nothing but yourself.' "[72]

The Gospel allusions here are significant. They take us full circle to where we began with this history. Back in the late nineteenth century, French medical doctors like Charcot and Bernheim argued that the reason Lourdes produced such dramatic healings—far more impressive than anyone had ever seen in a secular hospital setting—had to do with the fact that there was no more potent source of faith than God, no more potent theater for healing than a religious shrine. In our own time, we are seeing similar kinds of arguments. The study of the placebo effect, it is said, brings us back to bigger questions about the healing power of faith more generally: faith in doctors, faith in medical treatments, but especially faith in divinity. Religion, it has been suggested, is a system that, to a believer, can produce a supercharged placebo effect—and now we have the scientific evidence to prove it. Clergy across various Christian denominations are taking notice of these new arguments. One Catholic observer of the current scene talks about attending a meeting that included a panel of several chaplains from a major U.S. children's hospital:

> After a review of recent scientific studies linking prayer, faith and physical health, one of the speakers gasped, "When I hear such excit-

> ing news, all I can say is: 'Wow.' " As it turned out, he had more to say, mostly about how clergy now had objective data supporting their role in the "health care team." While I know this particular Lutheran minister to be sincere, conscientious and pastorally astute, I still couldn't help wondering if faith as "wellness technique" was really what Luther had in mind while writing his Commentary on Galatians.[73]

Coda

Today, debate about the nature and efficacy of the placebo effect continues apace, with skeptics attempting to assimilate placebo responses to other factors such as spontaneous remission and "regression to the mean," and advocates continuing to assert the reality of this effect.[74] In some very recent twists, functional magnetic resonance imaging (fMRI) studies have been recruited in an effort to visualize the beneficial effects of placebo treatments on brain functioning.

From an historical and cultural perspective, however, the real question is: What is the meaning of the continuing intensity of this debate? Why are we so inclined today to reconfigure our long-standing interest in the magic of positive thinking into a new interest in the magic of placebos? Part of the answer has to do with the way in which doing so allows us to buttress our beliefs in positive thinking and self-healing with hard data from laboratories, epidemiological research, and more. Connecting positivity and placebos allows us to conceive new ways of studying the healing power of faith experimentally, even using brain imagining technologies. While it is not clear how one could put a positive thinker in an fMRI machine and learn anything significant about self-healing, it is relatively easy to imagine one group of subjects being given fake morphine and then comparing what happens inside their brains to those of subjects who have been given real morphine.[75]

The search for scientific respectability, especially through the brain sciences, is thus part of the answer as to why the placebo effect has become the touchstone of our time for talk about the healing power of faith and positive attitude. I think it is not, however, the only reason. Having spent some years now talking with scientists, journalists, and nonspecialists

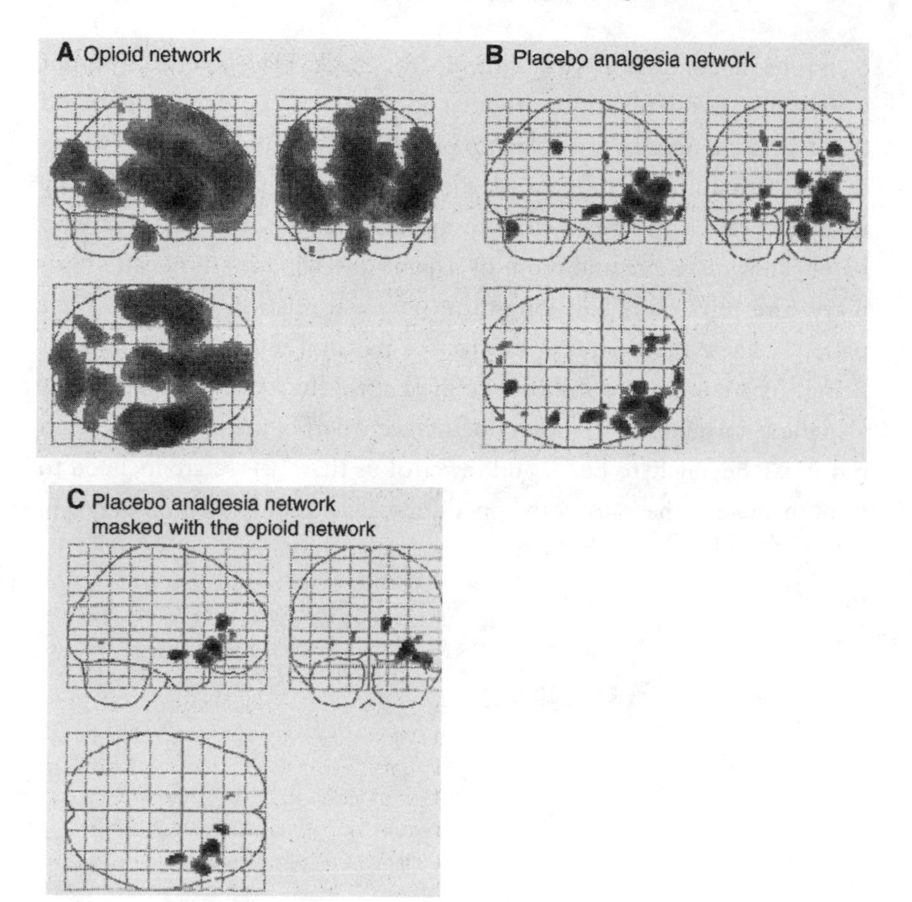

PET scan from the 2002 *Science* article by Predrag Petrovic and colleagues comparing the effects on the brain when patients are given the drug remifentanil (an opioid) as opposed to when they receive a placebo (an inactive solution). Image C shows the common areas of the brain that are activated in response to both the active treatment and the placebo treatment. Figure 2a from Predrag Petrovic et al., *Science* 295 (March 1, 2002): 1737–40. *Reprinted with permission from AAAS*

about the placebo effect, I have become convinced that our interests in the real healing power of the placebo effect have roots also in a kind of discontent with modern medicine that is distinctly different from that of the 1970s. When positive thinking first medicalized itself under Norman Cousins, advocates for a new holistic medicine were concerned, above all, with finding ways to empower patients to resist a style of mainstream clinical care that was widely perceived as arrogant and paternalistic. In our own time, in contrast, we want our doctors back. Placebo effects do not work—at least, not usually—without a doctor's personal involvement; and that fact is important to people. No one wants to return to the era where the doctor's authority was unquestioned. But in a time dominated by managed care and fifteen-minute appointments with clinicians who barely know us, more and more of us are nostalgic for a time when primary care physicians had long-term personal relationships with their patients, knew their names, and took a pastoral interest in their well-being. As a culture, we still believe in positive thinking, but many of us also think it would be nice if our doctors were part of the formula. It turns out it can be lonely to be ill, and fewer of us than before are inclined to want to undergo the journey to hoped-for wellness completely on our own.

Chapter Four

Broken by Modern Life

Future shock [is] the shattering stress and disorientation that we induce in individuals by subjecting them to too much change in too short a time.

—*Alvin Toffler,* Future Shock, *1970*

Norman Vincent Peale was the spokesman sine qua non for the power of positive thinking. But in 1957, in one of his syndicated columns, he struck an unaccustomed new note. Alongside the need to develop one's capacity for positive thinking, he told his readers, was the need to do something about a growing problem of modern society: the problem of "stress." What did this unfamiliar word mean? Peale helped his readers understand by telling them a little story:

> A friend of mine is a pilot for one of the great airlines. He flies a big DC-6 and, when I fly with him, I enjoy the privilege of spending some time with him in the cockpit.
>
> On one of these trips he was discussing the use of engine power. He stated: "If I use my full engine power for long periods, these engines will lose their efficiency. That is to say, stress isn't good for them." And he went on to describe how he uses every bit of his 2500 horsepower per engine to lift his mighty plane into the air at take-off. Then, as he begins to climb, he cuts back to 1800 and, when he reaches cruising level, he drops it still further to 1200.

> "If I ran those engines at 2500 for more than two minutes, it would cause strain in them which, in time, might injure them permanently. I have great power at my command in this plane, but I must know how to use it, how to conserve it."

Peale's point, of course, was that what was true for engines was true for people too. But he allowed his pilot friend to sum things up for him:

> "Each of us as people come to times in our lives when we need all the horsepower we can pour on. But if we use it constantly and don't know how to ease off, even God can't keep us in physical health. We just can't permit ourselves to go through life tearing ourselves to pieces by strain and stress."[1]

Horsepower, engines, strain, and stress—this is vocabulary from the world of engineering, a world we have not previously encountered. The presence of this vocabulary signals the fact that we are now working with a new kind of mind-body narrative, one that I call "broken by modern life." When we tell stories based on this narrative template, we are no longer the passive creatures who are healed in response to the suggestions of powerful authority figures ("the power of suggestion"); nor are we people who find ourselves physically healed when we have the courage to face and confess our secret sins and traumas ("the body that speaks"); nor, again, are we the fortunate possessors of wells of energy we can tap to heal ourselves ("the power of positive thinking"). Instead, we are complex, self-regulating machines who must husband our energies properly if we are not to risk permanently damaging ourselves.

Stress is the central scientific concept at the heart of this new narrative of mind-body medicine; but that is not all it is. Stress is also, for countless millions of people, an experience, and a very visceral one at that. More particularly, stress is an experience that particularly affects *us*—modern people, living fast-paced, modern lives: "Stress may be worst killer of the modern era"; "Modern Life includes toll exacted in terms of stress"; "Stress: modern man's silent enemy"; "Premature aging the result of modern stress." The headlines echo across the postwar period and into our own time.[2] Why is stress our peculiar burden? Because, the answer goes,

Time magazine cover: stress as the bane of modern life (June 6, 1983).

modern life makes demands on us that are fundamentally unnatural. Life may have been challenging for our ancestors, but at least they lived in some kind of harmony with their environment. When confronted with a threat, they could rise to meet it; when it was mastered, they could rest again. For us, we think, things are different:

> The queue at the supermarket checkout and the traffic jam can become the fangs of the sabre-toothed tiger and when confronted by these threats we respond just as if that tiger were there—by activating our caveman stress response. Having activated our body for an immediate physical response, there is often no need or opportunity for physical action! . . . We cannot fight the queue; we cannot run away from it either. So we become impatient and irritated; we become angry; we fume![3]

Before stress: "Exhausted" by modern life

The experience of fuming in the supermarket checkout queue is so clearly stressful to us that we are tempted to suppose that, while the scientific study of stress may have a history, stress itself does not. People have always known stress, we think; it is just that we know more stress than earlier generations. But is this the last word on the matter? The concept of stress as we know it has only been in existence for about sixty years, and there is at least suggestive evidence that, in the period before people learned to think of themselves as stressed, they experienced the woes of life, even of modern life, rather differently. Back in late-nineteenth-century America, for example, when modern life became too much, people did not stew in a chemistry of frustration; rather, they fell victim to deficiencies in their "nerve force." They took to their beds; they became incapable of dealing with bright lights, excessive noise, or surprises; and they suffered from respiratory disorders like asthma, hot and cold flashes, skin rashes, and sick stomachs. Their problems had less to do with being chronically overwrought than with being chronically exhausted.

In the 1880s, New York neurologist George Beard gave diagnostic specificity to this general visceral experience by naming it "neurasthenia" (a term actually borrowed from an older clinical tradition). Neurasthenia was a disorder, Beard said, that was caused by a malfunctioning of the nervous system, and if left untreated, it could result in collapse. Cases could be found throughout the civilized world, but Americans in particular were prone to neurasthenia because of the fast pace of American life, the harsh American climate, and the advent of various modern novelties that had changed life generally in this country:

> The causes of American nervousness are complicated, but are not beyond analysis: First of all modern civilization. The phrase modern civilization is used with emphasis, for civilization alone does not cause nervousness. The Greeks were certainly civilized, but they were not nervous, and in the Greek language there is no word for that term. . . . The modern differ from the ancient civilizations mainly in these five elements—steam power, the periodical press, the telegraph, the sciences, and the mental activity of women. When civilization, plus these five factors, invades any nation, it must carry nervousness and nervous diseases along with it.[4]

Why would all of these modern developments result in nervousness? Beard found his understandings and explanations in the modern world itself. Friendly with the American inventor Thomas Alva Edison, Beard drew on Edison's work on the electric lightbulb as a rich source of metaphors for thinking about the dynamics of the exhausted human body. In 1881, just three years after Edison's public demonstration of the electric light, Beard had this to say:

> Edison's electric light is now sufficiently advanced in an experimental direction to give us the best possible illustration of the effects of modern civilization on the nervous system. . . . [W]hen new functions are interposed in the circuit, as modern civilization is constantly requiring us to do, there comes a period, sooner or later, varying in different individuals, and at different times of life, when the amount of force is insufficient to keep all the lamps actively burning; those that are weakest go out entirely, or, as more frequently happens, burn faint and feebly—they do not expire, but give an insufficient and unstable light—this is the philosophy of modern nervousness.[5]

What could be done for people with neurasthenia? Beard, who took his electrical metaphor seriously, advocated direct application of electric currents to the bodies of neurasthenic patients. However, a more widely practiced treatment for neurasthenia, at least for women, was total bed rest. Developed by the American neurologist S. Weir Mitchell, the so-called rest cure for neurasthenia involved taking women away from their families, sending them to bed, forbidding them to move, denying them access to any visitors, and feeding them a diet designed to build up their "fat and blood" reserves.[6]

For men, the cure for neurasthenia was different. A respite from the pressures of modern life was also advised, but in this instance it was accomplished by sending the patients out into primitive, natural surroundings. As Mitchell put it, when it came to men, "The surest remedy for the ills of civilized life is to be found in some form of return to barbarism."[7] The young Teddy Roosevelt was among the cases of male neurasthenia in this time who was prescribed a regime involving vigorous excursions into the wilderness. It is tempting to speculate that, were it not for those youthful experiences, the American national parks system, established some years later under Roosevelt's presidency, might never have come into being.

But for every cure, there were two new cases. By the early twentieth century, the incidence of neurasthenia had reached what appeared to be epidemic proportions. A writer for the *North American Review*, H. Addington Bruce, observed in 1908, "On every street, at every corner, we meet the neurasthenics."[8] A new group of elite, private-practice neurologists—so-called nerve doctors—emerged to treat the growing numbers of the infirm and the distressed in the wealthier classes, while people of more modest means tended to favor the tonics and pills that were sold by the patent-medicine industry as ways to calm and strengthen "bad nerves."

And then it all ended. Today, the neurasthenic diagnostic category has virtually vanished from our collective memory. Instead, we believe ourselves to be in the midst of an epidemic of stress. We have no difficulty seeing stress as the natural response of our species to the unnatural demands of modern life, and we find visceral confirmation for our beliefs on a daily basis. How did all this happen?[9]

Fighting like cats and dogs: Walter B. Cannon and the "fight or flight" response

Two men, above all, helped launch what we now understand as stress into the modern world. The first was the Harvard physiologist Walter B. Cannon, the second the Czech-born endocrinologist Hans Selye, who was a professor at the University of Montreal. In contrast to Beard, neither of these men was primarily a clinician. First and foremost, they were experimental physiologists. In other words, stress first came into our lives, not as a disease, not as a human experience, but as a discovery in the laboratory.

The relevant history here begins with Cannon. When he was still a student at Harvard, his mentor, Henry Pickering Bowditch, set him and another student, Albert Moser, the task of using newly developed X-ray technologies to study the passage of food through the digestive tract of laboratory animals. The work proved fruitful, and Cannon went on to use X rays to study peristalsis—the intestinal movements associated with digestion—about which at the time very little was known.[10]

It was in the course of these investigations of peristalsis that something unexpected happened. Cannon found that whenever his experimental animals—mostly cats—became distressed or enraged, their peristaltic activity was inhibited. This led to new questions: Why should distress or

Walter B. Cannon (1871–1945), at work in his laboratory at Harvard Medical School. *Courtesy of the Harvard Medical Library in the Francis A. Countway Library of Medicine*

rage inhibit peristalsis? What did emotions and digestion have to do with each other?

To find out, Cannon developed a protocol that involved putting cats in a safe cage, and then bringing in their "natural enemy"—a dog—to sniff and bark at them. The cat would become agitated (at least, sometimes), and Cannon would then draw blood from the animal and compare it with the blood of cats that had not been made to feel threatened.[11] Following a lead from a colleague, Daniel Roy Hoskins, who was working in a laboratory in Ohio, Cannon found that the blood of frightened cats always contained a certain hormone then called adrenin (today called adrenaline or epinephrine). At the time, it was known that adrenin was secreted by the adrenal gland (located on the kidney). It was also known that injecting adrenin into an animal resulted in certain physiological changes: increase in blood pressure and blood sugar levels, dilation of the pupils, piloerection (the "standing of hair on end"), and inhibition of digestion. Nothing in the understanding of the time, however, suggested any reason why all of these physiological changes should be associated with emotional excitation.[12]

In the end, it was a Darwinian perspective that made sense of everything for Cannon. He realized that all the physiological changes observed in a laboratory animal confronted with a "natural" adversary improved that animal's ability either to fight or flee from its enemy:

> The increase of blood sugar, the secretion of adrenin, the altered circulation and polycythemia in pain and emotional excitement have been interpreted . . . as biological adaptations to conditions in wild life which are likely to involve pain and emotional excitement, i.e. the necessities of fighting or flight.[13]

In "conditions in wild life," the animal, having mobilized its resources, either successfully masters the threat or fails to do so. If it fails, it ends up becoming prey. If it succeeds, its physiology settles back to a pre-emergency state. Cannon coined the term "homeostasis" for this process of dynamic regulation. Later, he showed that not just the adrenal gland but also specialized parts of the nervous system (now known as the sympathetic and parasympathetic systems) were centrally involved in such biochemical regulatory processes.[14]

All of this is famously part of the history of modern physiology. But from our point of view here, Cannon's next move was key. He noted that human beings have the same capacity as other animals for fight or flight reactions; we too have homeostatic systems to help us regulate the use of these reactions in response to environmental challenges and threats. In the modern era, however, Cannon suggested that life had become so fast paced, so uncertain, and consequently so anxiety provoking that many people went through their days as if they were cats faced with dogs perpetually barking at them. With their "emergency" responses thus chronically stimulated, there were few opportunities for homeostatic mechanisms to restore their physiologies to a resting state. Over time, therefore, modern human beings were prone to falling ill in what could be considered characteristically modern ways:

> The bodily changes in emotional excitement may be considered as an anticipatory of many . . . dangers. The forces of the organism are put on a war-footing. But if there is no war to be waged, if the emotion has its natural mobilizing effects on the viscera when there is nothing to be done, obviously the very system which functions to preserve constancy of conditions within us is then employed to upset that constancy. It is not surprising, therefore, that fear and worry and hate can lead to harmful and profoundly disturbing consequences.[15]

Cannon delivered these remarks as part of a lecture to colleagues in 1936, in which he urged medical practitioners to wake up to the fact that they were living in a new world of illness experience. Patients today were largely suffering not from old-style "plagues and pestilences" but from the "strains and stresses" of modern life; and orthodox medicine was largely failing to take the resulting disorders seriously. "Taught to deal with concrete and demonstrable bodily changes," he said reprovingly, "we are likely to minimize or neglect the influence of an emotional upset, or call the patient who complains of it 'neurotic,' perhaps tell him to 'go home and forget it,' and then be indifferent to the consequences."[16]

In fact, as Cannon's own laboratory work showed, it was clear that emotions were not just in the head; they were also in the body. Chronic emotional upset did not just cause people to feel unhappy; it could cause them to become physically ill. This was the lesson, Cannon said, that mainstream medicine needed to learn if it wanted to deal with the new realities of illness in the modern era.

Of rats and men: Hans Selye and the laboratory invention of "stress"

Cannon understood his research to be about the physiological basis and clinical effects of strong emotional experiences, but he did not use the word "stress," certainly not in the way we have come to use it today. It fell to Hans Selye to do that, and in so doing to transform Cannon's concepts of homeostasis and fight or flight into something new, something that involved emotion, to be sure, but that was not completely defined by it.

Selye was a Czech physician and biochemist, trained in Prague, who fled the Nazis in the early 1930s and finally found work in the Department of Endocrinology at the University of Montreal. His first position was as an assistant to the endocrinologist James B. Collip, who was looking for evidence of the existence of a new female sex hormone he had reason to believe might exist:

> I was to go out to the slaughterhouse with a large bucket and bring it back to him as rapidly as possible, filled with the ovaries of freshly slaughtered cows. . . . [Collip] made various extracts of this material and, as I was the youngest member of the academic staff, I was assigned the tedious task of injecting these into female rats and looking for any kind of change produced that could not be ascribed to one of the known hormones of the ovary.[17]

Selye duly carried out this task over a period of some months, but when it came time to perform the autopsies on the treated rats he was disappointed to see there had been no change in their sex organs. However, they did all suffer from a curious triad of symptoms: peptic ulcers in the stomach and upper intestine; enlarged adrenal glands; and shrunken immune tissues. Were these symptoms some sort of unexpected consequence of the bovine ovarian extract?

To find out, Selye began to test the effects of other organ extracts on the physiological functioning of the rats. He found that extracts of kidneys and spleens produced the same effects as extracts of ovaries. Then he had a new thought: perhaps what the rats were exhibiting was not a specific response to a specific agent but a nonspecific response to the trauma of having a noxious (and probably impure) agent injected into their bodies. He began to wonder if other kinds of trauma would result in the same outcome—and experimented by making life very unpleasant for many rats. Some were put on the roof of the medical building in the winter; some were put down in the heat of the boiler room; some underwent an operation in which their eyelids were sewn back and they were then exposed to brilliant lights; some were placed inside barrel-like, revolving treadmills powered by an electric motor that forced them into a state of complete exhaustion. "It gradually turned out," he said later, "that no matter what type of damage I inflicted on an experimental animal, if it survived long enough and the stressor was sufficiently strong, the typical combination would be produced: adrenal hyperactivity, lymphatic atrophy and peptic ulcers."[18] His early findings were published as a letter to the editor in the journal *Nature* in 1936.[19]

Hans Selye (1907–1982), photographed ca. 1962. © *Bettmann/Corbis*

What, however, did it all mean? In his memoirs, Selye talked about the way in which the rat experiments had put him in mind of an experience he had had as a young medical student in Prague. His teachers there had been very concerned that he learn to distinguish one disease from another, but he had been more struck by the fact that all of the patients he was asked to examine actually presented similar stereotypical symptoms, especially early on; they were all weak and listless, with similar facial expressions. At the time, he had called this "the syndrome of just being sick," and he asked his professors whether he might make this a special object of study. He later took a kind of pride in reporting how scornful they were of his ideas.

Because now, looking at his experimental rats, Selye felt vindicated: "Maybe all my injections merely damaged the rats, producing 'the syndrome of just being sick.' " He then pushed the logic of this conclusion still further: perhaps he had found here a common pathway for the emergence of a whole range of common but poorly understood disorders that most clinicians would not imagine had anything to do with one another. "I asked myself . . . why so many people suffer from heart disease, high blood pressure, arthritis, or mental disturbances. These are not completely stereotyped signs of all illness, yet they are so frequent that I could not help suspecting some non-specific common factor in their causation."[20]

By 1950, he had formally baptized this common factor "stress." He took the word from metallurgy, but it was, in fact, somewhat misapplied since in the world of engineering "stress" referred to the forces that act to deform or weaken metals; it did not refer to the resulting condition of the metals themselves.[21] By this time, too, Selye had further conceptualized the stress response as a nonspecific response of the body to any demand—what he called a "General Adaptation Syndrome." Drawing explicitly on Cannon's earlier work, Selye proposed that the response unfolded in three stages. In the first, the "Alarm Stage," the animal perceived itself as under threat and prepared for action, either fight or flight. Endocrine glands released hormones responsible for accelerating heartbeat and respiration, elevating blood sugar, increasing perspiration, dilating the pupils, and slowing digestion.

In the second, "Resistance Stage," the body repaired any damage caused by the alarm reaction, and all the physiological changes associated with that first reaction were reversed. The organism now also had increased resistance to the event that had originally caused its stress (what Selye would soon name the "stressor").

The third and final stage was only reached in those situations when a stressor persisted beyond the body's ability to respond effectively. Selye called this the "Exhaustion Stage." During this stage, mechanisms enabling adaptation began to show signs of exhaustion, planting seeds for the emergence of stress-linked diseases. In the most severe cases, the outcome of this stage was death.

The implications for humans in modern times, Selye thought, were clear. In a fast-changing world filled with stress, each of us had a choice: learn to adapt or risk illness or worse. "The secret of health and happiness," wrote Selye, "lies in successful adjustment to the ever-changing conditions on this globe; the penalties for failure in this great process of adaptation are disease and unhappiness."[22]

This was Selye's stress theory, but many of his colleagues in endocrinology and experimental physiology—including Cannon—reacted to it skeptically.[23] They suggested that Selye had exaggerated the uniformity of the responses seen in different experimental situations, had not adequately defined stress as a concept, and that, in any event, many of his experiments were highly artificial and had little if anything to say about pathophysiological processes seen in a clinical context.[24]

Instead of backing off or continuing to try to persuade his skeptical colleagues, however, Selye responded by looking outside physiology for allies and supporters. Over the years, he cultivated an extended audience for his stress theory in broad-based publications, lectures, general-interest articles, and more. He reached out to military psychiatrists concerned with the challenges facing soldiers in the new Cold War era; to restless psychosomatic medical clinicians looking for alternatives to the old Freudian nostrums; to general practitioners looking for new ways of making sense of patients with elusive symptoms; to special-interest lay groups (The Young Presidents' Club, the Million Dollar Round Table, the Maharishi Mahesh Yogi's International Meditation Society); and to ordinary people who read magazines like *The Readers' Digest*. Selye was especially appealing to people who knew they felt worried or unwell, but were perhaps no longer quite persuaded by the doctrine of bad nerves that had helped their parents and grandparents make sense of their experiences of malaise.[25]

The strategy worked. As early as 1956, one commentator opined that Selye's ideas had "permeated medical thinking and influenced medical research in every land, probably more rapidly and more intensely than any other theory of disease ever proposed."[26] By the 1970s, discussions of stress had become routine in advice columns, social analyses, self-help literature, and popular magazines.[27] In 1981 the Institute of Medicine estimated that some $35 million had been spent on stress research in 1979 alone, made note of the "thriving industry" of remedies and literature designed to help alleviate the problem, and pointed to the fact that many occupational-health disputes and legal cases now involved accusations of "excessive stress" on the job.[28] One of the most prominent cases of this sort happened in 1981, when the U.S. Professional Air Traffic Controllers Organization (PATCO) went on strike for the first time in its history, on the grounds that its controllers were being subjected to intolerable levels of stress. Researchers hired by the Federal Aviation Administration to investigate this had failed to find standard biochemical and behavioral indicators of stress, however, and Ronald Reagan, newly elected to the presidency, ended up firing all 11,359 of the striking workers.[29]

World War II and the militarization of stress

Selye's marketing efforts on behalf of stress were greatly aided by the fact that the idea resonated with broader clinical and social concerns of the time. One of the key catalysts in creating receptivity to his new vision of illness in the modern world was World War II. Many felt that this war had marked a significant change in the nature of the challenges faced by troops in combat settings, that they were forced to grapple with unprecedented assaults on their minds and bodies. In 1945, two military psychiatrists working on the North African front, Lt. Col. Roy R. Grinker and Maj. John P. Spiegel, clarified some of the ways in which this was so in their milestone study *Men under Stress*.[30] In this book, they described their analysis of bomber pilots, men whose mission it was to drop bombs from planes over enemy territory. The work was brutal, in part, people said, because it was so unnatural; it put the pilot (in the words of a later analyst) "in an environment for which his evolutionary history could not possibly have fitted him."[31] The cockpit in which he sat was tiny and cramped and prone to extremes of low temperature and lack of oxygen at high altitudes. If anything went wrong with his equipment, he was helpless. Danger was all around him. And he was by himself, isolated from all comrades. Military planners knew that bomber pilots could stand only so many sorties before they would start to unravel. Grinker and Spiegel's book was an attempt to see what specific factors caused some men "under stress" to reach their breaking point faster than others.

In spite of the title of their book, Grinker and Spiegel were not building on Selye's laboratory-based, physiological vision of "stress." For Grinker and Spiegel, stress simply connoted hardship or affliction (a colloquial definition of the word dating back to the Middle Ages); they wrote about their military patients from a doggedly psychoanalytic perspective. More specifically, they conceived their work as a study of the ways in which preexisting neurotic vulnerabilities in combatants could be brought to the surface by the hardships of war, resulting in breakdown, maladaptation, or neurosis. In their words, "The question . . . is: How much did his previous personality and how much did the stress he experienced contribute to his reaction?"[32] Franz Alexander, Flanders Dunbar, and Freud made up prominent parts of Grinker and Spiegel's bibliography; Selye was not even cited.

In the postwar era, however, Selye was able to persuade several military psychiatrists that there was in fact an important link between the laboratory work on what he was calling stress and the larger concerns they had about the new challenges facing fighting men in modern warfare. They would get further, he convinced them, if they started to pay attention, not just to the unconscious conflicts supposedly experienced by their soldiers but also to their adrenaline levels. And in the end, many were persuaded. In fact, in the 1950s and 1960s the military did more than any other group to raise the profile of the idea of stress, extending its apparent explanatory reach out of the laboratory and into the real world. By the 1970s, as many as a third of all researchers doing stress research were based in U.S. military institutions: the Walter Reed Army Institute of Research and the Stress and Hypertension Clinic of the Naval Gun Factory in Washington, D.C.; the Military Stress Laboratory of the U.S. Army; the Naval Medical Research Unit in Bethesda, Maryland; and the Stress Medicine Division of the Naval Health Research Center in San Diego, California.[33]

Nevertheless, from the beginning these researchers faced formidable challenges in adapting a research program originally developed for laboratory rats to military concerns. It was not just a matter of developing protocols that could be applied to people—that kind of effort had already begun. It was more that, in order to be of much use for military planning, experimenters needed to find experimental stressors that would at least approximate the kinds of stresses that a soldier could expect to experience in a true combat situation. Mental arithmetic tests and mild electric shocks were not going to satisfy that imperative. In fact, in one remarkable study from 1962, a group of particularly resourceful researchers subjected soldiers to a series of highly realistic simulated crisis situations without telling them they were participating in research. One scenario involved leading the men to believe they were trapped in an airplane that was about to make a crash landing; another involved isolating the soldiers in an area they were told was about to be subjected to artillery attack; still another involved isolating them in an area that they were told had been accidentally exposed to radiation. Some were also left in a building that was supposedly in imminent danger of being engulfed in a forest fire (they could see the smoke); and others were led to believe that, because of an error they had made, a comrade of theirs had been seriously injured in some undisclosed underground location. The researchers investigated the effects of such experiences on circulating hormone levels, and they explored how well the men continued to perform on a range of crucial tasks.[34]

It was clearly not practical, however, to do studies like these on a regular basis. If the insights and understandings of the laboratory were going to be adapted to the real-world interests of the military, therefore, researchers realized they would have to find a different way forward. In this context, some focused on the fact that Selye's model of stress had emphasized that what broke an organism usually was not a single, sharp experience of stress, but rather cumulative experiences over time. If that was the case, then potentially the whole world of everyday life could be conceived as a laboratory capable of producing data relevant to military concerns. After all, in these modern times, everyone suffered varying degrees of stress. Was it possible that, over time, this sort of stress could add up to effects just as noxious as any experienced by a soldier on the battlefield? From this perspective, some began to argue that the differences between military stress and civilian stress had been exaggerated:

"many conditions of ordinary life—for example, marriage, growing up, facing school exams, and being ill—could produce effects comparable to those of combat."[35]

A key technology in reinforcing this new perspective was a questionnaire called the Social Readjustment Rating Scale (SRRS). Designed in the mid-1960s by psychiatrists Thomas H. Holmes and Richard H. Rahe (significantly, Rahe had also been a U.S. Navy commander), the test aimed to give researchers a way to quantify the cumulative effects of stress on individuals over a period of a year or more. At the heart of the test was a scale that identified forty-three more or less common stressful life events—events that required adaptation to change—and rank ordered them from 0 to 100. "Death of a spouse" was first on the list, worth a full 100 "life change units" (LCUs); "taking out a big mortgage" was midway down, worth 31 LCUs; and going through Christmas holidays was near the bottom of the list, worth a mere 12 LCUs.[36] (As a matter of interest, the original scores had been determined by making marriage the fixed midpoint on the scale, and then asking groups of people to rank the other stressors in relation to that fixed point.) What mattered, according to this line of research, was not whether the changes in question had a distressful or joyful meaning for the individual, but the amount of adaptation each demanded. If a person's cumulative LCU score was above 200 over the course of a year, he or she was considered to be at significant risk for certain diseases; over 300, and a person was considered to be gravely at risk.[37]

For the American poet Robert Sward, this way of thinking about the challenges of life worked on him—literally—like a strange poem:

To be married and moderately unhappy
is less stressful than to be unmarried
and male and over 30.

To be happily married counts for "0" points.
If your spouse dies that counts for 100 points.
63 for going to jail. 73 for divorce.

Divorce is more stressful than imprisonment.
Getting married is 3 points more stressful

SOCIAL READJUSTMENT RATING SCALE

Rank	Life event	Mean value
1	Death of spouse	100
2	Divorce	73
3	Marital separation	65
4	Jail term	63
5	Death of close family member	63
6	Personal injury or illness	53
7	Marriage	50
8	Fired at work	47
9	Marital reconciliation	45
10	Retirement	45
11	Change in health of family member	44
12	Pregnancy	40
13	Sex difficulties	39
14	Gain of new family member	39
15	Business readjustment	39
16	Change in financial state	38
17	Death of close friend	37
18	Change to different line of work	36
19	Change in number of arguments with spouse	35
20	Mortgage over $10,000	31
21	Foreclosure of mortgage or loan	30
22	Change in responsibilities at work	29
23	Son or daughter leaving home	29
24	Trouble with in-laws	29
25	Outstanding personal achievement	28
26	Wife begin or stop work	26
27	Begin or end school	26
28	Change in living conditions	25
29	Revision of personal habits	24
30	Trouble with boss	23
31	Change in work hours or conditions	20
32	Change in residence	20
33	Change in schools	20
34	Change in recreation	19
35	Change in church activities	19
36	Change in social activities	18
37	Mortgage or loan less than $10,000	17
38	Change in sleeping habits	16
39	Change in number of family get-togethers	15
40	Change in eating habits	15
41	Vacation	13
42	Christmas	12
43	Minor violations of the law	11

The original 1967 "Social Readjustment Rating Scale," by Thomas H. Holmes and Richard H. Rahe. The text accompanying the introduction of this scale in the journal *Psychosomatic Research* noted that "these events pertain to major areas of dynamic significance in the social structure of the American way of life." Critics would later point to gender and other kinds of biases in the original list, and changes were made accordingly. From "The Social Readjustment Rating Scale," *Psychosomatic Research* 11:213–18. *© Elsevier, 1967*

than being fired. Marital reconciliation (45 points)
and retirement (also 45 points)
are only half as stressful as
the death of your spouse.

Minor violations of the law: 11 points.
Trouble with the boss: 23. Christmas: 12. But
sexual difficulties are less stressful
than pregnancy (40 points versus 39).
A mortgage over $10,000 is worse
than a son (or daughter) leaving home.
Trouble with your in-laws is as stressful
as "outstanding personal achievement"
which is only slightly more stressful
than if "wife begins or stops work." . . .

Conclusion: With 25 points or more, "you probably
will feel better if you reduce your stress."[38]

Postwar prosperity and the overburdened executive

Even though the stresses of modern life spared no one, there still might be differences worth registering in the ways that the burdens of stress were distributed across the population. Did the contented manual laborer really know the same kind of stress as his harried boss? Did the housewife, spared the burden of needing to draw a salary, really understand the kinds of stresses her husband confronted on a daily basis? The answers seemed self-evident.

Nevertheless, it wasn't long before the laboratory seemed to confirm them. In 1958, American psychologist Joseph Brady published a suggestively titled article called "Stress in 'Executive' Monkeys." The article described a study in which pairs of laboratory monkeys were subjected to cycles of electric shock. Each monkey had access to identical-looking levers, but only one of the levers was actually functional: when pressed, it deactivated the electric circuits. The monkey with access to the working lever was called the "executive" monkey and was harnessed to his companion in such a way that, when he pressed his lever in a timely fashion, both animals were able to avoid shock. Brady reported how, over a period of several days, the "executive" animal in each round of the experiment was more likely to develop gastric ulcers than his passive counterpart (even though both received exactly the same number of shocks). The conclusion seemed obvious: the role of executive decision-maker produced far more stress, and stress-linked disease, than did the role of passive follower.[39]

Later, in the early 1970s, psychologist Jay Weiss challenged this conclusion, suggesting that in fact Brady's "executive" monkeys suffered from stress-linked disease not because they had to make all the decisions, but because the experimental setup actually denied them any sense of real control over their fate.[40] By that time, however, the idea of a link between stress and the executive or managerial lifestyle had become deeply embedded in American popular culture. It confirmed what everyone thought they knew, namely that life was tough at the top.

Indeed, the Brady study had been published at the height of the Eisenhower era, when many were convinced for other reasons that all was not well. This was, of course, an era of unprecedented prosperity. After World War II, the United States had consolidated its position as the world's wealthiest nation. A growing number of middle-class Americans were holding what began to be called "white-collar" jobs, working as corporate middle managers, office workers, salespeople, service employees, and teachers. Automobile production quadrupled annually between 1945 and 1955. Low mortgage rates (designed particularly to make home ownership affordable for returning serviceman) fueled a housing boom. Many of the houses in question were built in the new suburbs, where they encouraged a lifestyle organized around the nuclear family and material comfort.

Nevertheless, for many Americans these developments had produced less happiness than might have been expected. The 1950s was not just an era of television and two-car garages; it was also, as many historians have observed, the "age of anxiety."[41] Some of the reasons for the anxiety were obvious: the bomb, fears of a communist invasion, and widespread beliefs that democracy and freedom were being put at risk by spies within American society. Other reasons for anxiety, though, seemed to have more to do with pressures associated with the new prosperous lifestyles themselves. The prosperity itself had been partially achieved through a new conservative approach to gender roles. Women who had developed a taste for employment during the war years now found themselves under great pressure to return to the home in order to make more jobs available to returning soldiers.[42] In the 1950s, growing numbers of these women were diagnosed by doctors as deeply anxious in their own right, and were offered various remedies for their woes. These included psychotherapy designed to help them adjust to their roles as wives and mothers, and also new medications known as minor tranquilizers which helped take the edge off their most intense feelings. (The minor tranquilizers were later dubbed as "mother's little helpers," after the derisive Rolling Stones song of that name).[43]

At the same time, it was clear to many that America's menfolk were also in trouble. These were the years that saw the coining of terms like the "rat race" and "the tread mill," phrases that brought attention to the dehumanizing costs of maintaining middle-class lifestyles. New forms of popular literature and film began to glorify rebellion from the conventions of the time. In 1951, sociologist C. Wright Mills published an influential study called *White Collar: The American Middle Classes*, in which he

Image used in a 1964 advertisement for the tranquilizing medication Deprol, produced by Wallace Laboratories (today incorporated into MedPointe Pharmaceutics); the image here clearly identifies anxious housewives as a target for treatment, and the accompanying text promises to reduce "symptoms of depression and associated anxiety." *Printed with permission from MedPointe Pharmaceuticals.*

identified a new class of workers—the middle managers—and worried aloud about their values and psychology.[44] In 1955, Sloan Wilson's novel *The Man in the Gray Flannel Suit* (made into a film starring Gregory Peck the following year) gave vivid expression to what life might be like for these kinds of workers on the ground. The novel tells the story of a young married couple with three children who live miserable, martini-soaked lives in a Connecticut suburb, while Tom, the husband—who is also a World War II veteran—struggles to climb the corporate ladder in his Manhattan firm. Eventually, he finds the courage to rebel against his situation.[45]

Finally, in 1956, sociologist William Whyte offered what felt to many like an incisive analysis of men in gray flannel suits like Tom. In his book *The Organization Man*, he wrote:

> This book is about the organization man. If the term is vague, it is because I can think of no other way to describe the people I am talking about. They are not the workers, nor are they the white-collar people in the usual, clerk sense of the word. These people only work for The Organization. The ones I am talking about belong to it as well. They are the ones of our middle class who have left home, spiritually as well as physically, to take the vows of organization life, and it is they who are the mind and soul of our great self-perpetuating institutions.[46]

Now, the consensus already existed that the wives of these organization men suffered from anxiety. But how was one to describe the situations of the men themselves? Increasingly, people were inclined to draw on the manly, military language of stress to help the men give voice to and seek remedies for what ailed them. Some of these men probably ended up taking the same minor tranquilizers as their wives. But being able to think of themselves as stressed meant they could seek help without having to feel stigmatized by a feminized diagnostic label. Indeed, in 1957 the pharmaceutical company Charles C. Pfizer made a short industrial film, *The Relaxed Wife*, that aimed explicitly to encourage doctors to think of the stressed businessman as a target for treatment with minor tranquilizers. Filled with strategically humorous images (e.g., a businessman with a head literally about to explode under pressure), lots of talk about the tensions of corporate life, and the importance of learning to relax, the film managed to promote its anxiolytic product, Antarax, without using the word "anxiety" once.[47]

Men coping badly: the rise of the Type A personality

In the 1950s and early 1960s, virtually everyone involved in the larger debates about the health costs of stress agreed that stress was a cumulative experience. Pressure built up inside a man's head until he was at risk of exploding. The burdens of decision making ate up the insides of the overstressed executive until he developed ulcers. The Social Readjustment Rating Scale was fundamentally grounded in this understanding of stress as cumulative.

And yet by the late 1960s and 1970s, considerable evidence had accumulated to suggest that this approach to stress might be fundamentally misguided. Rather than responding in a uniform way, different people seemed instead to respond to the same stressors, both in the lab and in life, in a range of ways. Some people seemed to be remarkably resilient even in the face of considerable provocation, while others crumpled under what would appear to be only modest provocation. Still others appeared to remain composed but showed impaired performance.

Results like these led some stress researchers—notably University of California, Berkeley psychologist Richard Lazarus—to insist that the entire engineering approach to stress as cumulative burden or load was misleading. Unlike machines, Lazarus argued, humans (and to some extent laboratory animals) bring a set of cognitive skills to the circumstances in which they find themselves. When they face a novel situation or potential stressor, people engage first in appraisal behavior to decide if they are indeed threatened. Then they review their options and determine if they have the capacity to cope with the stressor. Stress occurs, according to Lazarus, only when people conclude that they face a threat they are incapable of mastering.[48] In this sense, Selye and other physiologically oriented stress researchers had missed a key mediating variable that stood between a world full of dangers and the biological stress response: the human capacity for creative coping.

Potentially, this insight changed everything. If stress was redefined as a thing that was less objectively imposed on a person by the world and more a function of how effectively that person coped with the world, then this raised a new question: Was it possible that there were certain types of people who consistently coped less well than others with the challenges of life? In other words, were there some people who were characterologically predisposed to manufacture unhealthy levels of stress for themselves?

In the 1960s, an increasing number of people answered this question with a yes. Indeed, a consensus began to emerge in the field that there existed a particular kind of man (and originally it was always a man) who suffered from dangerously high levels of stress, especially occupational stress, not because his workload was objectively so demanding but because he placed endless demands on himself. Competitive, obsessed with deadlines, always in a hurry, men like this were soon given a name: Type A personalities. At the time, the evidence seemed clear that, unless they changed their ways, such men risked a tragic end: premature death from heart failure.

The discovery of the Type A personality and his apparent vulnerability to heart attack had roots in a far more broadly based set of public health concerns in the postwar period. This was a period of sharply rising levels of coronary heart disease in the American population. Once seen as a relatively rare disease in the United States and Europe, by the second half of the twentieth century, heart disease had come to be called the silent epidemic of the times, responsible for some 30 percent of deaths in industrialized countries—the largest single cause of death from any disease.[49]

Why this sharp rise in the incidence of heart disease? In the late 1940s and 1950s a number of epidemiological efforts were launched to try to find out. Most notable among them was the Framingham Heart Study, which was the first to confirm a link between smoking and heart disease and to show that physical activity reduced one's risk of heart disease. Also important was the Seven Countries Study, which compared heart disease rates in the United States, Yugoslavia, Italy, Greece, Finland, the Netherlands, and Japan and provided evidence linking heart disease to diets high in saturated fat.[50]

In spite of such studies, some researchers felt that epidemiological investigations should be extended beyond diet and exercise to embrace the possibility that some kinds of occupations also might put people at risk. The reason had to do with a skewing of the epidemiological data; while heart disease was on the rise in general, its incidence was particularly high among males who self-identified as white-collar managerial types—so much so that some cardiologists had taken to calling coronary heart disease the executive disease.

Why might white-collar executives be more at risk than others? In the mid-1950s, San Francisco cardiologists Meyer Friedman and Ray Rosenman began to make the link to stress. As they later told the story, they had noticed that the front edges of the chairs in their waiting rooms were worn down by fidgeting, impatient cardiac patients from that occupational sector.[51] Was there a link between the chronic impatience of these men and heart disease, they wondered? To see what others thought, they sent a survey around to a hundred and fifty San Francisco businessmen and a hundred general practitioners, asking them the same question: Were there any personality traits they believed characterized people who had heart attacks? From the list of options on the questionnaire, more than 70 percent of the businessmen and a majority of the internists picked "excessive competitive drive and meeting deadlines."[52]

This was clinical intuition. The question was, could it be validated objectively in a clinical study? To find out, Friedman and Rosenman and two colleagues studied forty accountants for six months: three months before April 15 (the date when tax returns must be filed in the United States) and three months after that date. Every two weeks, these physicians measured their subjects' cholesterol levels and the speed with which their blood clotted (rapid clotting was considered a risk factor for heart disease). Sure enough, by late March and early April, both had increased significantly. Indeed, the researchers reported that the accountants' blood "began clotting at a dangerously accelerated rate." In May and June, though, both cholesterol and clotting returned to normal levels. Since nothing else had changed substantially in the lives of these men, the conclusion seemed to follow that the stress of meeting the deadline for filing tax returns had increased the accountants' risk factors for heart disease.[53]

Put another way, this study of accountants suggested that a certain kind of pressured working environment could affect heart functioning. It did not, however, imply that by virtue of their personalities, particular groups of professionals, such as accountants, were chronically at risk for heart disease. After all, when the deadline for the delivery of tax returns passed, the accountants' health improved. A year later, however, Friedman and Rosenman went much further. In a study titled "Association of Specific Overt Behavior Pattern with Blood and Cardiovascular Findings," they compared three groups of men, identified simply as groups A, B, and C. Group A consisted of eighty-three men who all manifested "an intense, sustained drive for achievement" and were known to be perpetually "involved in competition and deadlines, both at work and in their avocations." Group B also consisted of eighty-three men, but these were supposed to be the exact opposite of group A—relaxed, easygoing, and noncompetitive. The third group, C, was a kind of control group: it consisted of "46 unemployed blind men" who manifested "a chronic state of insecurity and anxiety." The stunning punch line of this study was: "Chronic coronary artery disease was seven times more frequent in group A than in group B or group C."[54]

Emboldened by these results, Friedman and Rosenman continued with a much more ambitious project, the Western Collaborative Group Study, which attempted to see if Type A behavior did not just correlate with heart disease but could actually predict it. This new study used a standardized structured interview technique to screen some 3,100 healthy volunteers for Type A behavior tendencies and then proceeded to track them over time. After eight and a half years, the men who had originally been classified as exhibiting a Type A behavior pattern turned out to be twice as likely to develop heart disease as the men who had been judged originally to exhibit a Type B behavior pattern. The results were not as spectacular as those reported in the original correlation study, but they were statistically significant—and they supported the hypothesis.[55] In 1981, a review panel from the U.S. National Heart, Lung, and Blood Institute thus formally announced that it judged the evidence to be conclusive: Type A behavior pattern was an independent risk factor for coronary heart disease in middle-aged U.S. citizens in industrialized geographical areas.[56]

If this was the case, what was to be done? Would medicine now insist that the corporate lifestyle of prosperous postwar America was no less a health hazard than smoking? Would epidemiologists now demand fundamental changes in the working conditions of American executives? If people were willing to blame modern life in general for a range of stress-linked disorders, did it not follow that they should also be willing to blame the corporate world for its role in encouraging behaviors that increased employees' risk of heart disease and premature death? A few people moved cautiously in this direction, noting that the "20th-century Western milieu" had "probably increased the prevalence of [Type A behavior] . . . if only by offering special rewards to those who can perform more competitively, aggressively, and rapidly than others."[57]

In the end, however, the discussion about Type A behavior and heart disease failed to connect decisively to the larger social and political debates of the time about the price of prosperity in postwar America. Instead, the literature increasingly adopted a strategy that involved accommodation to existing workplace values. In part, this happened because it was felt that Type A personalities themselves were not fundamentally interested in changing—they stood to lose too much. In 1978, Rosenman wrote that doctors should take care to reassure their patients that "alternative behavior [to the classic Type A pattern] is not a threat to their socioeconomic well-being." On the contrary, by being willing to introduce small stress-defusing rituals into their existing routines, they were likely actually "to accomplish more with less strain" and to live longer, healthier lives.[58] At the same time, very little was asked of management: "if possible, a working environment should be created that will reduce noise, telephone interruptions, a chronically littered desk, etc.—all factors that elicit urgency and irritation."[59]

By the mid-1970s, the Type A personality had become a fixture of American culture. The media had played its part here, with hard-to-ignore headlines like "Heart Attack Personality: Will Success Kill 'Type A' Man?"; "You May Make a Killing. Personality a Heart Attack Sign"; "Stress No. 1 Coronary Factor? Type B Better off in a Type A World"; "Rushing Your Life Away with 'Type A' Behavior."[60] Some historians have argued that, as a cultural trope, Type A functioned as a kind of caricature of ambitious, competitive masculinity, and indeed it is perfectly possible that some people took a perverse pride in discovering evidence for Type A traits in themselves.[61] In their 1974 self-help book *Type A Behavior and Your Heart*,

Friedman and Rosenman even provided a handy quiz to help people in their efforts on this front:

> Do you have a habit of (a) explosively accentuating key words in your ordinary speech . . . and (b) finishing your sentences in a burst of speed?
>
> Do you always move, walk, eat rapidly?
>
> Do you get unduly irritated at delay—when the car in front of you seems to slow you up, when you have to wait in line, or wait to be seated in a restaurant?
>
> Do you often try to do two things at once?
>
> Do you almost always feel vaguely guilty when you relax and do absolutely nothing for several days (even several hours)?
>
> If you answered YES to any of these questions, you may be falling into the dangerous TYPE A pattern. You need to read this book, take the complete test and change your behavior to prolong your life.[62]

In addition to the kinds of homey changes that Friedman and Rosenman were recommending in these years—coffee breaks, fewer telephone interruptions, etc.—some people also began to promote a more scientific-seeming approach to stress reduction: relaxation training, the mastery of methods that were supposed to quickly produce a decrease in stress on an as-needed basis that in no way otherwise interfered with a person's busy lifestyle. In these years, the progressive relaxation training method developed in the 1930s by the American psychologist Edmund Jacobson became reconceptualized as a means of stress reduction and enjoyed renewed popularity.[63] By the end of the 1970s, Harvard cardiologist Herbert Benson's quick-and-easy method for evoking what he called a "relaxation response" was also winning large numbers of converts; I have more to say about Benson's work in chapter six.

Through much of the 1970s, the relaxation technique that had the most cachet—and also, in the eyes of many, the most clinical promise—was biofeedback. In the 1970s, some saw biofeedback as a sober clinical extension of operant conditioning methods originally developed in the animal laboratory, while others viewed it as a training technology that might help people achieve their "supermind." Either way, all agreed that it was likely to be good for reducing stress.

How would it do this? As the name suggests, biofeedback was (and is) an attempt to extend the scope of conscious control over physiological processes of which people are normally unaware. The basic method involves using instruments to measure and display changes in particular physiological processes, and then to train individuals to use the resulting information to self-regulate their bodies in ways that would normally be impossible for them.

One of the important early pieces of evidence that people might be able to do this had come from the work of psychologists Neal Miller and Leo DiCara. These researchers had shown that laboratory animals temporarily paralysed with curare (a poison traditionally used to coat blow darts in South America) could be trained using operant conditioning techniques (reward and punishment) to alter their visceral physiology in specific target directions. Responding to reinforcers, rats learned to alter their heart rates and even their brain waves; and dogs learned to alter their intestinal activity and salivation levels.[64]

For human subjects independently motivated to achieve the target results, the feedback process itself generally took the place of reinforcement. Hooked up to the right instruments, it was said, people could gain self-control over a huge array of disorders, from incontinence to asthma to epilepsy. Always at the top of the list of promised benefits, however, was the idea of relaxation on tap—of using biofeedback to turn off the stress response at will.

One satisfied customer of a biofeedback training program designed for this purpose told a journalist what the experience had been like:

> He responded to an ad for the new Biofeedback Training Center, 645 N. Michigan Av., and spent a month taking the $225 course. In each of the ten sessions, he sat in a comfortable womb chair in a soundproof room. Electrodes strapped to his forehead enabled him to listen to a sound feedback of his frontalis muscle tension and watch a series of lights that also illustrated his tension level.
>
> When he clenched his teeth, the lights shot from green to yellow

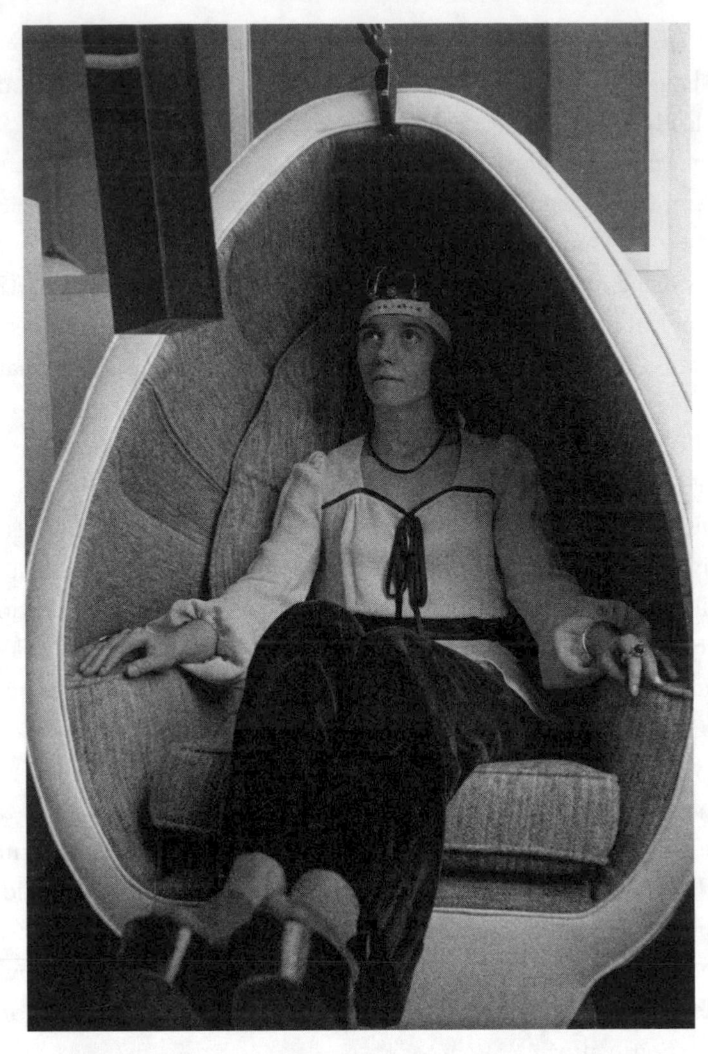

Image from 1970s of the new technology of biofeedback, highlighting its usefulness as a technology to combat stress. The original caption accompanying this image read: "CHICAGO, October 9, 1975. Got a headache? . . . Relax as Joanna Spilman is doing while watching [the] monitor of a biofeedback machine enabling her to see and hear tension levels in her body. The band around her head picks up the electrical nerve activity on the frontalis muscle (a bandlike muscle running across the forehead). Electrical energy is measured with sensitive instruments then 'feedback' to the feedback device so the person can see when his frontalis is relaxed or tense. With practice, he or she can learn the 'feeling of relaxation' and promote the state at will." © *Bettmann/Corbis*

> to red, indicating maximum tension. As he learned to relax, the lights changed one by one until they were solid green. He learned the "feeling" of relaxation, which, with regular practice, he can now produce at will.

The journalist went on to note, pointedly, that the customer in question had gotten into the habit of turning on the "feeling" during crucial business meetings, because he had found that doing so allowed him to "concentrate better on the problems at hand."[65]

By the 1980s, the warning messages about Type A behavior had been absorbed. The link between the workaholic lifestyle and heart disease seemed about as solidly established as one could hope for; the relaxation industry was in full swing . . . and then things began to unravel. Skepticism had already been growing in some circles about certain aspects of the methodology used in the Western Collaborative Group Study (particularly aspects of the interviewing protocol), but they remained largely under the public radar screen until 1988. In that year, an article appeared in the *New England Journal of Medicine* that reported the results of a follow-up study of the subjects who had participated in the Western Collaborative Group Study. When assessed at eight and a half years as already mentioned, it appeared that men with Type A traits were twice as likely to die of heart attacks as the so-called Type B men. On a longer timescale, however, the trend had failed to hold—in fact, the follow-up data showed heart disease to be modestly lower among Type A than among Type B men. "This apparent advantage associated with Type A behavior is surprising," the authors of the study commented wryly, "and needs confirmation, but the results do indicate that patients with CHD and a Type A behavior pattern are not at increased risk for subsequent CHD mortality."[66]

Other researchers insisted that these surprising results did not mean there was no link whatsoever between certain personality types and heart disease. New analyses now began to suggest that just two traits associated with the original Type A behavior pattern—but originally not given much emphasis—did indeed conduce to heart attack: hostility and cynicism.[67] With this new twist, however, no longer are we unambiguously in a world in which people are broken by the stress, self-imposed or otherwise, of modern life. Instead, we are in a new narrative world about the effects of poor interpersonal relationships and social isolation on health (this is the subject of the next chapter, "Healing Ties"). By the late 1980s, new best-sellers like *The Trusting Heart: Great News about Type A Behavior* and *Anger Kills* began to crowd out older favorites like *Type A Behavior and Your Heart*.[68] In 2000, after late-night television host David Letterman underwent surgery for heart disease, a piece appeared in the popular press entitled "David Letterman's Cynical Heart." The writer, Robert Wright, put the matter to his readers straight:

> What is wrong with David Letterman's heart? The official reason for his quintuple bypass last month was atherosclerosis—clogged arteries. Some observers, such as *People* magazine, go deeper in search of the explanation, citing "Type A," workaholic behavior. But I submit—with the support of actual scientific evidence—that what's wrong with Letterman's heart is the same thing that is wrong with his TV show: excessive cynicism.[69]

Stress and the immune system

Before the end of the 1970s, it was generally felt that so-called stress-linked disorders (on the rise) and infectious disorders (on the decline) were distinct categories of illness. Throughout the early period of research on stress and disease, there seemed to be no obvious way in which stress could play any role in a person's susceptibility to, or recovery from, an infectious disease. What allowed a person to fight off infection was his or her immune system—a highly sophisticated biochemical and cellular defense system capable of recognizing and then protecting the body from a range of "foreign invaders" (bacteria or viruses). Medical science was persuaded that the immune system performed these functions in a wholly autonomous fashion, unaffected by input from the nervous system or any other part of the body. Nothing could deter it from its self-appointed path. Indeed, it had been found that one could take the relevant cells and substances out of the host body and put them in a petri dish with some toxin or foreign agent, and they would still move smoothly into action, just as they would under normal (in vivo) conditions.

During the late 1970s, however, a series of challenges emerged to this consensus understanding of the immune system. New evidence began to suggest that the immune system was not in fact wholly independent of other systems in the body after all, and that, in particular, its functions could be modulated in a range of previously unsuspected ways by inputs from the nervous system. In one key experiment (also discussed in chapter three), psychologist Robert Ader and immunologist Nicholas Cohen showed that the immune system of rats could be modified using a classical Pavlovian conditioning method: a pairing of saccharine water with a powerful immune-suppressing drug, cyclophosphamide. After conditioning, when rats drank the saccharine water alone, their immune systems continued to show signs of reduced functioning, and the animals continued to die. Of course, such findings made no sense in the framework of the time, but "as a psychologist," as Ader wryly recalled some years later, "I did not know there were no connections between the brain and the immune system."[70]

In the late 1970s and 1980s, conditioning studies such as these from behavioral science began to be supplemented by other discoveries of functional connections between the immune system and the nervous system. For example, it was found that lymphocytes—part of the cellular defense system of the immune system—produce peptides that had previously been believed to reside only in the brain, suggesting that the two systems might be in some kind of chemical communication with each other. Other work began turning up evidence of hard-wired connections between the nervous system, thymus gland, spleen, lymph nodes, and bone marrow (all important sites of the immune system).[71] It was Robert Ader who proposed that all this new work in fact represented the dawning of a new discipline, which in 1980 he christened "psychoneuroimmunology," or PNI.[72]

From the beginning, people who were drawn to the new field made links to stress and stress research. Part of the reason for this had to do with timing: PNI emerged as a new discipline just as AIDS—a disease that undermined immune functioning—was emerging as a terrifying new epidemic in the West.

The thinking here was clear: if the immune system could be influenced by information coming from the nervous system in general, then potentially, at least, it could be undermined by stress. This possibility in turn suggested that stress might be a terrible risk for the highly vulnerable immune systems of HIV-positive or AIDS patients, and should be avoided at all costs. But how could stress be avoided? The fateful diagnosis itself imposed an immediate stress on patients. The anthropologist Emily Martin tells the story of how one of her subjects found out his lover had tested HIV positive and then faced the difficult task of getting himself tested. At the clinic, he and the other people in the waiting room were made to watch a film about AIDS that vividly showed how the HIV virus destroys cells in the human immune system. He told Martin how he had fled from the clinic, feeling persuaded that viewing the film would itself undermine his immune functioning.[73]

Fear of AIDS diagnosis itself was a chronic stressor for AIDS patients. But so too, many felt, was what some researchers began to call "the stress of social homophobia," the enormous hostility toward gay lifestyles manifest in much of mainstream American culture. In his book *And the Band Played On*, AIDS activist Randy Shilts quoted American evangelist Jerry Falwell, saying of the thousands who were dying at the time of AIDS, "When you violate moral, health, and hygiene laws, you reap the whirlwind. You cannot shake your fist in God's face and get by with it."[74] In an article from 1986, the psychologists George Solomon and Lydia Temoshok were blunt about the potential public health stakes of this kind of societal intolerance: "To the extent that these stresses are immunosuppressive, then they may be involved in the increased vulnerability of these groups to AIDS."[75] In other words, a generation of young men was dying not just from the HIV virus but from the chronic stress of living with the rejection and disdain of society.

The AIDS-patient community, for its part, seemed to have little trouble incorporating the new ideas about stress and the immune system into a series of politically impassioned stories about what was truly killing them. One early (1986) grassroots volume for and by AIDS patients, for example, had this to say:

> We believe that the AIDS virus particularly strikes individuals and groups who have been isolated by the dominant culture. . . . It is this isolation, often internalized as self-hatred or lack of self-acceptance,

> which allows the AIDS virus to begin to incubate once it has entered the system. . . . It is no coincidence that the rise of AIDS has to a large extent coincided with the recent upsurge of right-wing political and religious repression of gays.[76]

In the late 1980s, the first AIDS drugs arrived—not drugs that could cure AIDS, but drugs like AZT that did seem to significantly slow its progression. These developments were followed in 1995 by the first true pharmacological treatments for the disease—the so-called protease inhibitors. With a growing focus on these developments, both medical research and grassroots patient activism began to shift toward building on the promise of conventional molecular approaches to fighting the disease.

Nevertheless, the early years of the AIDS crisis created a lasting legacy for the stories our culture would tell about the meaning of stress. They taught us that the modern era had not escaped the age-old plagues of infectious diseases after all; and they also taught us that the well-recognized syndrome of modern life, stress, had a reach that had not previously been suspected. We learned that, by undermining immune function, stress could compromise our ability to defend ourselves against infection. In this sense, AIDS taught us that the modern era had been doubly caught out.

By the second half of the 1990s, research on stress and the immune system had begun to focus on other vulnerable populations. Did the so-called Gulf War Syndrome result from the fact that highly stressed American soldiers, already facing combat during the first Iraq war in 1991, were then given multiple powerful vaccines—as many of 24 different kinds—that their stressed-out immune systems were unable to properly defend against? In an aging population, were there particular stressors—such as mourning the death of a spouse, or caring for a spouse with Alzheimer's—that undermined already vulnerable immune systems?[77] What about cancer? Did stress affect the ability of the immune system to fight tumors in ways that accelerated the progress of the disease? Certainly, many cancer patients began to think so. In a series of interviews I conducted with cancer patients some years ago, I found they spoke frankly of their fears of stress. "If I find a way to handle it [stress]," one woman told me, "then I know it's not drilling holes in my brain, and I'm not wandering around worrying about it all day; 'that's depressing my immune system, that's maybe making me, et cetera, et cetera.' " And another said: "I'm trying not to, you know, to put myself under the pressure of worrying too much . . . because I don't want, you know, to have this product [cortisol] in my body. . . . If I get angry, [I] go out and scream somewhere so I can get rid of it."[78]

With virtually any and every disease now potentially at risk of being made worse by stress, our "broken by modern life" lament has reached a new level of intensity. Perhaps this helps explain, at least in part, why in the 1980s and 1990s we see increasing attention paid to two new narratives of mind-body medicine. In its own way, each of these new narratives functions as a counterpoint to the "broken by modern life" lament. Even though we live in a harsh, fast-paced, and unnatural world filled with stress, these two new narratives tell us that there are ways to heal, to recover balance, to boost immunity, and to increase well-being. In the final two chapters of this book, I explore where these new narratives came from and how they function in our culture.

Chapter Five

Healing Ties

Loneliness is a health hazard brought about by people losing the sense of community and family intimacy that has marked every other age.

—*Rev. Dr. Gordon Moyes, sermon given on September 23, 2001*[1]

In 1998, the American magazine *Newsweek* published an article titled "Is Love the Best Drug?" The article included an inset with the following questions:

Do you feel isolated?

Women who said yes, were three-and-a-half times more likely to die of breast, ovarian or uterine cancer over a 17-year period.

Does your wife show you her love?

Men who said no suffered 50 percent more angina over a five-year period than those who said yes.

Are you close to your parents?

Male medical students who said no were more likely to develop cancer or mental illness years later.

Do you feel loved?

Heart patients who felt the least loved had 50 percent more arterial damage than those who felt the most loved.

Do you have a confidant?

Unmarried heart patients who said no were three times more likely to die within five years.

Do you live alone?

Heart-attack survivors who said yes were more than twice as likely to die within a year.[2]

The numbers here tell a story, a story based on a narrative template I call "healing ties." I might almost as readily have called it "fractured ties," because it is a narrative that skirts the border between melancholy and hope. A product of the same postwar American preoccupations that gave birth to "broken by modern life," this narrative agrees that we do indeed have reason to lament the high price of modern life. What we should be lamenting, however, is less the relentless pace of our lifestyles and more our loss of community. The stress we experience, this narrative says, is real. But it is less a cause of our woes and more a symptom of the deeper problem we face: unnatural disconnection from our fellow human beings. "People who lack social support," the *Newsweek* article explained, "tend to stew in stress hormones all the time, not just when they're counting down by 17s." Over time, this kind of chronic loneliness can—literally—break our hearts. That conclusion in turn leads to a simple prescription: to get well again, we need to recommit to community and intimacy; we need to recommit to one another. Human ties are healing ties.

Of course, on some level we did not need mind-body medicine to tell us that there was something wrong with modern social relations. A critical literature going back more than a century had long lamented the loss of close-knit lifestyles and rural values and called attention to the social and moral dangers supposed to result from ever more people moving to the impersonal, crowded cities. At the close of the nineteenth-century, the French sociologist Émile Durkheim had presented statistical evidence that suicide was more common in societies that had recently experienced a breakdown in social norms—what Durkheim called "anomie" (a Greek word meaning "absence of norms or rules")—than it was in more stable societies. Within individual societies, he found that suicide rates were also relatively higher among groups of people who were less socially integrated and regulated by collective societal norms: unmarried people (as opposed to married people), people living in urban settings (as opposed to people living in rural villages), and Protestants (as opposed to Catholics).[3]

In postwar America, some sociologists adapted Durkheim's concept of anomie to describe other costs they believed were associated with living in an increasingly mobile and rootless society. Anomie, often partially conflated with its more psychologically minded cousin "social alienation," was thought to be able to help explain not only changing patterns of suicide but rising levels of juvenile delinquency, sexual deviancy, criminality, and mental illness.[4]

Roseto, Pennsylvania: A Shangri-la in the heart of modern America

Few of the sociologists who pointed to the dangers of anomie and social alienation appear to have imagined, however, that these dangers might be medical as well as psychological and social. All this changed in the 1960s, with the discovery of a little town in central Pennsylvania, Roseto, that appeared to be enviably immune from the great epidemic illness of the age: heart disease. The paper trail here begins with an article published in the *Journal of the American Medical Association* in 1964 called "Unusually Low Incidence of Death from Myocardial Infarction." In it, epidemiologist Clark Stout and his colleagues offered data from death certificates culled from hospital and other town records from Roseto that suggested something quite remarkable about this town. In comparison with its two neighboring towns, Bangor and Nazareth, Roseto seemed to be a virtual haven from heart disease, despite being serviced by the same water supply and hospital as its neighbors. Nationally, it was known at the time, the frequency of death from heart disease rises with age. In Roseto, however, it was near zero for men aged fifty-five to sixty-four. For men over sixty-five, the rate was half the national average.[5]

One of the authors of this 1964 study was a psychiatrist named Stewart Wolf who had a long track record of interest in psychosomatics and stress. Wolf was particularly known in psychosomatic circles for his investigations in the 1940s of a patient named Tom with a gastric fistula that had allowed his fascinated doctors to peer directly at the man's digestive

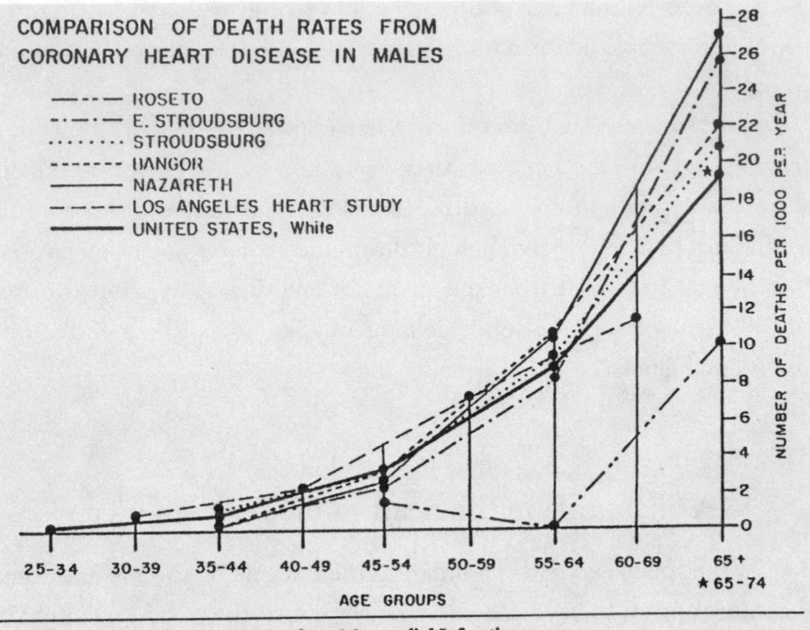

Comparison of Death Rates in Men from Myocardial Infarction

A Shangri-la in the heart of modern America? The rates of death from heart attacks in Roseto, Pennsylvania, as compared to neighboring towns (Nazareth and Bangor) and also as compared to the national norm, from Wolf and Bruhn, *The Power of Clan*, copyright © 1993 by Transaction Publishers. *Reprinted by permission of the publisher*

processes and track the ways in which they were affected by changes in his emotional state.[6] Wolf, who spent his summers in eastern Pennsylvania, had first learned about Roseto from a local physician who told him he had rarely, if ever, seen any Rosetan patient under the age of fifty die from myocardial infarction. These were the years when such large epidemiological studies as the Framingham Heart Study and Seven Countries Study were just beginning to pinpoint smoking and a high-fat diet as major risk factors for heart disease.[7] If the emerging findings of these other studies were comprehensive, then Roseto should have been a case study in unhealthy living; for many Rosetans were overweight, and many smoked. People regularly feasted on generous portions of Italian sausages and meatballs that had been fried in lard rather than olive oil (which the Rosetans could not afford to import from Italy). As for occupational stress, another emerging risk factor for heart disease, they seemed to have that too: the average man from Roseto worked long and tedious hours at the local quarry; the average woman worked equally long shifts at the local blouse factory.

Wolf was determined to get to the bottom of what he would later call "the Roseto paradox." Teaming up with a sociologist named John Bruhn, he reviewed the history of the town, conducted extensive interviews with the citizens, observed their lifestyle over years, and concluded that there was one key thing that set Roseto apart from its neighbors. Unlike Bangor and Nazareth, Roseto had been settled early in the previous century by immigrants from a poor town in southern Italy who had traveled together from the Old Country to build a new life for themselves. On arriving in Pennsylvania, the newcomers "were forced by snobbish neighbors"—towns settled previously by immigrants from England and Wales—"to look out entirely for themselves, to support one another for survival and to form their own enclave."[8]

And therein, Wolf and Bruhn believed, lay the unwitting source of the residents' good health. Denied the opportunity to assimilate, the newly immigrated Rosetans created an oasis of Old World values and customs in the heart of a rapidly modernizing America. Multigenerational homes were the norm, life revolved around the Catholic Church and various civic organizations, and everyone (in the words of Wolf and Bruhn) "radiated a kind of joyous team spirit as they celebrated religious festivals and family landmarks."[9] People up and down the street all ate the same sorts of meals on a reassuringly regular rhythm (pasta on Tuesdays, fish on Fridays, etc.), and standoffishness or one-upmanship among neighbors was explicitly frowned upon. "The local priest emphasized that when preoccupation with earning money exceeded the unmarked boundary, it became the basis for social rejection." For this reason, "despite the affluence of many, there was no atmosphere of 'keeping up with the Joneses' in Roseto."[10] All these factors together, Wolf and Bruhn concluded, were the secret to the town's heart-healthiness. Roseto was living proof, as they put it, of "the power of clan."[11]

Even as those investigations were in progress, however, there were signs that all was not well in Roseto's Old World paradise. Interviews with younger Rosetans indicated that many in the new generation actually harbored more "typically American" dreams of a bigger house and fancier lifestyle, and increasingly resented and resisted the old ways. As one of these young people put it:

> There is very little excitement. No industry, which is the reason for young college graduates to abandon this town. There is no place to get ahead. Other than the mills, there is nothing a person can do for a living. . . . The children in Roseto have a chance for a good education, but it is hard to live in Roseto if you acquire a specialized education. All Rosetans have a higher goal, but old people want to keep things the way they are used to.[12]

In 1963, reading the writing on the wall, Wolf made a bold prediction: if the new generation of Rosetans did indeed abandon the ways of the "old people," Roseto would cease to be a haven from heart disease. It did not take many years before he had the opportunity to test his hypothesis. Young people started marrying non-Italians, leaving the Church, moving into new suburban houses, joining country clubs, and taking up golf; some women even joined Weight Watchers! In the words of one observer, "It seemed like a capsulized, accelerated fulfillment of the American dream."[13]

Was it worth it? Some of the younger Rosetans themselves admitted that their new, more typically American lifestyle came at a price. "Everything is modern here, very nice," one housewife told Wolf a little wistfully. "I have everything I need, except people. When we lived in town, the neighbors were always in my kitchen and I was always in theirs. We talked. We knew what was going on there and there was always someone around to help you and keep you from feeling lonely. I miss that," she concluded, "but I guess I will never go back."[14]

What were the health implications of giving up neighborly chats in the kitchens for the sake of a swimming pool in one's backyard and a second car? In 1971, Roseto found out: the first heart attack death of a person younger than forty-five years old occurred in the town. And things got worse. In spite of new efforts by townspeople to cut down on smoking and fat consumption, coronary heart disease more than doubled through the 1970s, hypertension tripled, and there was a substantial increase in strokes. By the end of the decade, the number of fatal heart attacks in the town had risen to the national average. Wolf's prediction, it seemed clear, had come tragically true.[15]

Japanese hearts, East and West

Roseto was and remains, for many, nothing less than a morality play that really happened; an apparent vivid illustration both of the power of healing ties and of the medical dangers involved in fracturing those ties. It was not, however, the only such case study to function in this way. Sharing the spotlight in those years with Roseto was another story, one about the Japanese and what happens to them when they come to the United States and adopt Western ways.

The origins of this second "healing ties" story lie in the mid-1960s, when University of California, Berkeley sociologist Leonard Syme became involved in a project with epidemiologist Reuel Stallones to study rates of coronary heart disease and stroke among Japanese immigrants to Hawaii and California. At the time, the Japanese were of great interest to epidemiologists because of evidence showing they had the lowest overall rates of heart disease and lived on average longer than any other nation group on the planet. Most epidemiologists at the time attributed the Japanese's longevity to their low-fat diet, and the dietary theory was the starting point for Stallones's work. Would immigrant Japanese-Americans adopt a more high-fat, Western-style diet as they assimilated? And would such a transition affect Japanese-American rates of heart disease?

Syme was interested in something else. He had done earlier research which indicated that men who had either recently moved or changed jobs were more likely to suffer a heart attack than men who had done neither. This was especially true when a man was forced to relocate from a rural setting (e.g., laboring on a farm) to an urban office job. A high degree of "cultural mobility," Syme had concluded, was also a risk factor for heart disease, and so he wondered how far, independent of any changes in their diet, the stress of immigration and acculturation would affect the health of Japanese-American men.

In the end, as Syme later recalled, he and Stallones "were both surprised by the findings." The Japanese men who immigrated to California turned out to have rates of coronary heart disease five times higher than the rates in Japan, while immigrants to Hawaii had rates intermediate between those in Japan and California. This heart disease "gradient" could not be explained by diet, Stallones reluctantly agreed (changes in diet in both the Hawaiian and the Californian groups were similar.) But neither could it be explained by Syme's "cultural mobility" hypothesis (after all, both groups had emigrated and experienced dislocation). Confronted with these anomalous findings, Syme did what any sensible researcher would do: "I assigned a doctoral student to figure out what was going on."[16]

The student was an Australian named Michael Marmot, who completed his dissertation on the heart disease gradient in Japanese immigrants in 1975. Marmot's explanation for the gradient focused neither on diet nor on cultural mobility but on the degree to which Japanese migrants were able to maintain their traditional culture. It turned out, he said, that the "most traditional" Japanese-Americans living in California had coronary heart disease prevalence no higher than what had been observed in Japan. In contrast, "the group that was most acculturated to Western culture had a three- to five-fold excess in CHD prevalence."[17]

What did traditional Japanese culture offer that might account for these health benefits? Marmot's answer was: a close-knit community, that is, a community that provided its members with a great deal of stress-reducing emotional and social support. If Marmot was right, it began to look as if the Japanese lived longer than any other group on the planet not just because they ate a healthy diet but because they, perhaps more systematically than other countries, had developed a culture that had learned to exploit the power of healing ties.

The Alameda County study: with a little help from my friends

His dissertation behind him, Marmot went to England to begin a new project on risk factors for heart disease in civil servants, leaving his advisor Leonard Syme to continue to brood on the issues he had raised. As Syme recalled his thinking at that time:

> What does it mean to say "Western ways" versus "Traditional ways"?
>
> I went to Japan several times, interviewed dozen and dozens of people, and I read many books to get some understanding of this but all I could get out of this work was that my Japanese informants thought that Americans were lonely. I challenged this observation many times but dozens of people said that anyone could easily see this loneliness when you saw so many Americans walking on the street, alone. "Alone on the street?" I said. "That's not evidence of loneliness," I said. People all shrugged at my naiveté. So I returned to Berkeley, this was in 1975, and found another doctoral student who agreed to work on this.[18]

This new student was Lisa Berkman, who had begun her career in the late 1960s on the streets of San Francisco working as an advocate and activist for uprooted and dislocated groups. As a graduate student in public health, she was then exposed to the sociological theory of "social networks," a way of thinking about communities as systems of diverse, interconnected relationships (first developed by the sociologist J. A. Barnes in 1954).[19] Based on her earlier advocacy work in San Francisco, Berkman was predisposed to believe that social network theory might well be relevant to public health concerns. There were anomalies in the public health epidemiological data that no one had satisfactorily explained, and they all cried out, she thought, for a theoretical perspective that was currently lacking. As she recalled in an interview with me:

> Len [Syme] kept sort of pushing, saying . . . "Look at Seventh Day Adventists [who live longer than many other groups], look at Japanese-Americans, look at widows [who often die sooner than would have otherwise been actuarily expected] . . . what is it that explains their differences in health?"[20]

To begin to find out, Berkman agreed to undertake a dissertation project that involved analyzing an already existing set of data from a 1965 public health survey of a large group of residents of Alameda County, California. Because the survey had followed then-recent World Health Organization recommendations that "health" be defined by reference not just to physical but also to mental and social well-being, the participants in the 1965 survey—close to seven thousand people—had answered questions about their marital state, number of friends, and memberships in religious and voluntary organizations. In 1974, researchers had collected the nine-year mortality data on this group of people, but no one had thought to look for possible links between differences in the density of social networks and different mortality outcomes. Berkman's job now was to do just that.

What she found was stunning: in every age and sex category, people who in 1965 had reported the fewest social ties were, nine years later, up to three times more likely to have died than those who had reported the most social ties. This correlation held true even after factors such as socioeconomic status, smoking, alcohol use, obesity, physical activity, and use of preventive health services were accounted for.[21] Berkman's methodology was rigorous, and her results seemed unequivocal. Other epidemiological studies, looking at other populations, would later confirm them.[22] Soon there seemed to be no real room for doubt: having friends, being married, belonging to civic organizations, and belonging to a church were all conducive to healthier and longer life.

Community as "social support"

Why should this be so? What was the common active ingredient, responsible for the health effects of these different kinds of personal and community ties? In 1976 the epidemiologist Sidney Cobb gave this active ingredient a name: "social support." He defined social support as "information" that one is "loved, esteemed, and belongs to a network of mutual obligation." Possessing such information, he suggested, was stress reducing in ways that might explain the strong evidence for health benefits.[23] That same year, another epidemiologist, John Cassel, also talked about the apparent stress-reducing effects of what he called "social supports" in both animal communities and human societies. "These might be envisioned," he suggested, "as the protective factors buffering or cushioning the individual from the physiologic or psychologic consequence of exposure to the stressor situation."[24]

"Buffer" or "information"? From the moment it was coined in 1976, the term "social support" would be plagued by definitional ambiguity. If it was a buffer, then social support was probably best seen as something that people needed to keep on tap and dip into on a regular basis, especially in times of stress. If social support was information, however, then in effect it was simply a cognitive belief state, in which case it might not matter whether a person was really loved and esteemed or whether he or she had any real contact with others; all that might matter would be the person's subjective beliefs. This latter view came to be known as "perceived social support." Alongside these two very different understandings of social support was a third, more instrumental understanding, according to which the practical support of friends (a lift to a doctor's appointment, help in tax returns, advice on what to do about that unreliable boyfriend) became a way for people to reduce stress and extend their abilities to cope with life.

Even as people disagreed on how exactly to define social support, few questioned the strategy of looking for some common stress-reducing ingredient at work in people's experiences of family, friends, spouses, and community. The term itself stuck. Before 1976, only two articles had been published that contained the words "social support" in their title. Five years later, forty-three such articles had appeared; and five years after that, in 1986, eighty-three had appeared. By 1985, Cobb's 1976 article alone had been cited by other researchers more than four hundred times.[25]

Meanwhile, others outside the field did not wait for the definitional disputes to be resolved: whatever social support was, they wanted some of it. In 1984, the Mental Health Association of Onondaga County in Syracuse, New York, launched a public education campaign called "Friends can be good medicine."[26] *Self* magazine reviewed the evidence for the health benefits of social support in 1998, asking its readers "Have You Hugged Your Immune System Today?"[27] And in his best-selling book from 2000, *Bowling Alone*, political scientist Robert Putnam put the choices to his readers this way: "As a rough rule of thumb, if you belong to no groups but decide to join one, you cut your risk of dying over the next year in half. If you smoke and belong to no groups, it's a toss-up statistically whether you should stop smoking or start joining. These findings are in some ways heartening," he concluded. "It's easier to join a group than to lose weight, exercise regularly, or quit smoking."[28]

The politics of community; or what does hierarchy have to do with it?

But was that the end of the story? By the time Putnam was writing these words, something had happened to the original epidemiological argument linking better health to better community. Working away in London on his study of health differences among British civil servants, Michael Marmot was beginning to have some second thoughts. His results were showing that the lower a person's employment status within the civil service hierarchy, the greater the risk of death from diseases of all sorts. The chances of dying of a heart attack for those in the bottom tier of the civil service hierarchy were more than 2.5 times greater than for those in the top tier. Potentially, this made sense: people of lower socioeconomic status might not eat as healthy a diet, might smoke more, and so on. What was far more puzzling was this: across the entire civil service hierarchy, people in higher grades of employment were healthier than their immediate subordinates. As Marmot later put it, "The question is not why people at the bottom have worse health, but why social differences in health are spread across the whole of society."[29]

In attempting to answer this question, Marmot and his colleagues surveyed more than ten thousand civil servants. What was life like for them? The conclusion was that civil servants in lower-status jobs not only felt less socially supported, they also felt they had less control over their lives than people higher up the hierarchy; they felt competitive with and envious of others; and they reported experiencing more occupational stress in general.[30]

If Marmot's new findings were right, then perhaps it was necessary to rethink the then-accepted explanation for the longevity and good health of the Rosetans and traditional Japanese. Perhaps these people owed their good fortune less to the power of healing ties and more to the fact that they lived in communities where everyone was provided with a clear and secure social role, where overt displays of status were discouraged, and where conformity was encouraged. In Roseto, everyone ate pasta on the same days of the week. In Japan, adults might spend their entire lives working in the same corporation and follow a very predictable career trajectory. In the United States, in contrast, much of life was organized around a goal of continuing upward mobility, with resulting competitiveness, discontent, and stress. Such a lifestyle, according to this new way of seeing things, in the end had discernible effects on rates of heart disease and other kinds of morbidity and mortality.

This was a much more classically Durkheimian way of understanding the relationship between modern social values and health than the concept of social support had been, and it pointed toward a new kind of social critique: one that was less about community and more about capitalist economics. In the words of Ichiro Kawachi, director of the Harvard Center for Society and Health, and Harvard School of Public Health professor Bruce P. Kennedy, "striving after fame and fortune should come with a government health warning."[31]

If this was right, then clearly it was not enough simply to encourage people to join a bowling team. What was needed instead was an intervention that would be more political in nature and that would, over time, have the effect of reducing inequities in wealth and status across entire societies. In the words of Michael Marmot, "It is important to realize that many economic and fiscal policies may influence the social cohesion of a society. Those policies that increase income inequalities are also likely to create health inequalities.[32] We are no longer so clearly operating inside the narrative space created by "healing ties," but whether we are actually witnessing here the birth of a new mind-body medicine narrative—one that we could title perhaps "health and hierarchy" or "the Robin Hood effect"—is not yet wholly clear.

Community versus intimacy; or what's love got to do with it?

Even as these new debates within public health and epidemiology were playing out, the range of ways in which people told stories about healing ties was getting more complex. A new variation was emerging that was concerned less with the health benefits of community and more with the health benefits of something altogether more romantic: intimacy, or even (dare a scientist use that word?) love.

In 1977, University of Maryland psychiatrist and stress researcher

James J. Lynch published a book titled *The Broken Heart* that did dare to use the word. "There is a widespread belief in our modern culture that *love* is a word which has no meaning," Lynch declared. "This book was written to document the opposite point of view." Rising levels of heart disease had preoccupied the medical community for more than a generation, but until recently, Lynch went on, no one had drawn the obvious conclusion:

> Almost as soon as man discovered the existence and function of the heart, he recognized that it was influenced by human companionship and love. . . . Does common sense recognize something that scientists and physicians cannot see? Why do we continue to use phrases such as *broken heart, heartbroken, heartsick, heartless, sweetheart*? Why do people persist in the notion that their fellow men die of broken hearts when no such diagnoses ever appear on twentieth century death certificates?[33]

Indeed, there was good reason for the persistence of such notions, and, on some conflicted level (Lynch insisted), the medical profession knew it. In his book, he cited studies in the literature that pointed to: increased rates of heart attacks among bereaved spouses (a phenomenon actually known to doctors as "broken heart syndrome"); cases of sudden death following traumatic loss; evidence that premature death from coronary heart disease was more common in people who had lost a parent early in life; and his own research showing that the high, erratic heart rates of acutely ill and, sometimes, comatose patients in an intensive care unit could be stabilized simply by asking a nurse to take the patient's hand and speak to him or her in a caring way. The larger conclusion followed naturally: heart disease rates were on the rise nationally because the United States was sick of heart, making lifestyle choices that were resulting in toxic levels of loneliness.

To Lynch, it seemed clear even in the 1970s that the executive with Type A behavior was primarily at risk of heart attack not because he was so overworked but because his relentless workaholism had pushed away everyone who might otherwise have loved him:

> While the great majority of the individuals they examined were married, Friedman and Rosenman observed that the Type A personality

> engaged in a life-style that guaranteed a high degree of social isolation, not only from acquaintances but also from the immediate family. Often the Type A males were work addicts to the point where they grossly neglected their wives and children.[34]

Lynch's book hit a nerve, certainly in the popular press.[35] In short order, he was being interviewed on television talk shows and in popular magazines. *U.S. News & World Report* identified him as "an authority on the ways and misfortunes of the world's lonely people."[36] The medical community, it is true, was more ambivalent about him: the book was not widely reviewed in the professional medical literature. But one physician, reviewing the book for the *New York Times* in 1977, opined that *The Broken Heart* was "a moral tract disguised as science reporting" that offered "the wrong reasons for changing our way of life."[37]

Nevertheless, *The Broken Heart* caught a national mood, and through the 1980s lack of broader professional approval did not keep it from being seen as relevant to a wide range of questions, including: What to do about the social isolation of the disabled? What should be the priorities when it came to offering social services to the elderly? What were the real effects of divorce? And how should one understand the relationship between stress and social alienation?[38] In 1982, *The Broken Heart* was prominently featured in a long *New York Times Magazine* article titled "Alone: Yearning for Companionship in America." Journalist Louise Bernikow wrote there that, "Ordinary people cope with loneliness in ordinary ways," because they believe that "loneliness is to be lived with, like the weather. They don't think that they are in danger." But they are. "A broken heart," she concluded gravely, "is not a metaphor."[39]

Babies that fail to grow: lessons from an earlier era

Bernikow and others were worried above all about the health effects of ordinary loneliness on ordinary adults. But Lynch had also been deeply concerned about young children. He disapproved in particular of the feminist trend for mothers to work full-time out of the home. "Many children are, of necessity, raised in impersonal settings whose medical effects may not be fully apparent for decades," he lamented. "Parents can reduce contact with their children in the short term, and nothing may be apparent, but over the long run, serious emotional and physical problems may appear, even to the point of significantly shortening the child's life span some 30 to 40 years later."[40]

Warnings about the effects of parental—and particularly maternal—neglect on child development were of course not new. During the blitz in Great Britain in World War II, large numbers of children whose parents had been called away by the war effort or whose homes were believed to be at risk of bombing were placed in special residential war nurseries. Leading that effort was Anna Freud, recently escaped to England with her father, Sigmund. In collaboration with her companion and colleague Dorothy Burlingham, Anna did all she could to provide the very best of care, including encouraging parents to visit the children as much as possible. Nevertheless, it turned out that even the best institutional approach to child rearing was no substitute for a bond with a parent. In spite of receiving lots of attention and affection, the children began to lag behind in speech and other developmental milestones like toilet training. Some became listless or showed other signs of emotional dysfunction. Only when Anna Freud created artificial "family groups," in which three or four children were cared for exclusively by a single parental figure did things—slowly—improve.[41]

Wartime experiences like these helped solidify a consensus within psychoanalytic psychiatry in particular that any disruption to the mother-child bond, even when there was no overt abuse or neglect otherwise, could dramatically impair normal developmental processes. The man who, more than anyone else, helped to articulate this consensus was the British psychoanalyst John Bowlby. Commissioned in the late 1940s by the World Health Organization to write a policy report on the effects of maternal deprivation on child health, he summed up his conclusions in 1951 in a highly influential report titled "Maternal Care and Mental Health." It pulled no punches: "prolonged deprivation of a young child of maternal care may have grave and far reaching effects on his character . . . similar in form . . . to deprivation of vitamins in infancy."[42]

In the United States, another psychoanalytic psychiatrist, René Spitz, had provided an apparently vivid demonstration of the truth of Bowlby's proclamation with his famous 1945 studies—far more dramatic in their message than Anna Freud's work—documenting the emotional and physical regression of infants living in a foundling home. All of the children in this home, Spitz noted, received excellent physical care in a scrupulously hygienic environment. Good hygiene, it was believed, was the key to good health. At the same time, however, all of the children were given little if any individual love or attention. There was only one nurse for every eight infants, and physical contact was limited to brief feedings and diaper changes. Otherwise, each baby was kept isolated in a crib with sheets draped over the sides to prevent the spread of infection.

Instead of keeping the infants healthy, however, this regime had produced a collection of physically stunted, cognitively retarded, and emotionally damaged children. In Spitz's words, "By the end of the second year, the average of their developmental quotients stands at 45 per cent of the normal. This would be the level of the idiot. We continued to observe these children at longer intervals up to the age of four years . . . by that time, with a few exceptions, these children cannot sit, stand, walk, or talk."[43] When the children were between two and four years old, observers assumed they were seeing babies half that age. Finally, in spite of the scrupulous effort to maintain good hygiene, within two years 37 percent of the foundling children had died from infection.

Spitz then compared the fate of these foundling children to that of another group of children he had also followed. This second group of children lived in a prison nursery, where the environment was far dirtier and less regimented than in the foundling home. However, the mothers of these children (all of whom were prisoners) were permitted to spend a certain amount of time each day with their children, during which they lavished them with affection. The result? In spite of the dirt, in spite of the relative chaos of this prison nursery, not a single one of the second group of children succumbed to infection during the five-year period of Spitz's study. The mothers' love had proven a better deterrent to infection than the most conscientious of good hygiene practices.[44]

But even that was not all. There was also some evidence from this time that not just infants but also older children could be damaged by emotionally inconsistent or inadequate nurturing. In 1951, British nutritionist Elsie Widdowson described in the pages of *The Lancet* her studies of the effects of war rations on the health of children in Germany. Two

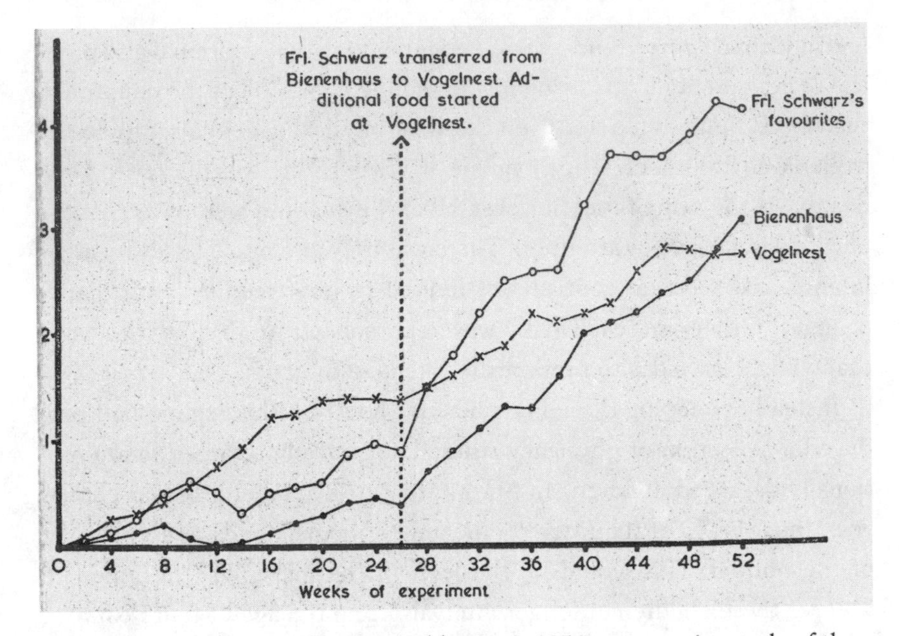

Growth curves of the orphans in Widdowson's 1951 comparative study of the effects of war rations on physical developments. Children in the orphanage called Vogelnest gained considerably more weight on the same war rations than children in the orphanage called Bienenhaus. The reason, Widdowson believed, was that Vogelnest was run by a loving matron, Fraulein Grün, and Bienenhaus was run by a cold and erratic matron, Fraulein Schwarz, who frequently terrorized and humiliated the children (often during mealtimes). Only a handful of favorites escaped her wrath. Halfway through the Widdowson study, Vogelnest was selected to receive extra war rations, and at the same time—wholly coincidentally—the loving matron left the orphanage and Fraulein Schwarz took her place (bringing her favorites with her). When this happened, the weight gain rates of the previously thriving children—even with the extra rations—plummeted, while weight gain levels at Bienenhaus improved. "Better is a dinner of herbs where love is than a stalled ox and hatred therewith," Widdowson concluded, with a quote from Proverbs. From *The Lancet,* 1 (1951): 316–18. *Reprinted with permission from Elsevier*

municipal orphanages in one of the British-occupied zones of Germany had been selected for the study. Each housed about fifty boys and girls between ages four and fourteen. For one year, Widdowson arranged for height and weight measurements of all the children to be taken every two weeks. During the first six months of the study, the children at both orphanages received exactly the same rations. During the second six months, however, Widdowson arranged for the children at one of the orphanages to be given a supplement of bread, jam, and orange juice. The simple goal was to see how much they benefited from the extra calories.

Examining her data at the end of the year, Widdowson was deeply puzzled. During the first six months, when they were eating identical rations, the children in one of the orphanages had gained considerably more weight and grown taller than the children in the other orphanage. This favored orphanage was then—by chance—also selected to receive the extra rations. One might therefore have expected these children, during the second six months of the study, to have grown even more than before. Instead, they suddenly began to languish, and by the end of the study their weight hardly exceeded that of the children who had remained for the entire year on standard war rations.

What was going on? The explanation, Widdowson finally concluded, lay not in the quantity of the food but in the quality of the matrons looking after the two orphanages. It turned out that the matron in charge of the orphanage where the children initially gained more weight was a motherly woman who was clearly fond of her charges. In contrast, the other orphanage was run, in Widdowson's words, by "an erratic martinet" who was "stern and forbidding, ruling the house with a will of iron. Children and staff lived in constant fear of her reprimands and criticisms, which were often quite unreasonable."[45] At the six-month point, when the orphanage run by the motherly matron was selected to receive extra rations, something unexpected happened. The motherly matron decided to leave her position, and the "erratic martinet" from the other orphanage took her place. Suddenly, the previously thriving children were deprived of affection and forced to live in fear. And the result—at least as Widdowson saw it—was that they stopped growing properly.

Spitz, Bowlby, and Widdowson wrote in a psychoanalytically and socially conservative era deeply concerned with the power of mothers in particular to act as a force for good or ill on their children. With the rise in the 1970s of a widespread revulsion against such so-called mother-blaming trends in psychiatry and medicine, the issue of the health-giving effects of mother love languished for a generation. In the 1990s, however, it became topical again. Following the collapse of Nicolae Ceausescu's dictatorship in Romania in 1989, Western doctors discovered 150,000 Romanian orphan babies who, like Spitz's foundling children, had been physically well cared for but left to languish without any affection or cuddling. Again like Spitz's foundling children, the Romanian babies had suffered enormously as a result. All were physically stunted and were both cognitively and emotionally impaired. Stress seemed to be at least part of the physiological explanation for this: researchers found that the children's cortisol levels were grossly skewed. At least one researcher of the Romanian tragedy was blunt about what she saw as the implications of her findings for her own society: "The consistent relationship between poor care and abnormal cortisol raises the question of what's happening to American children in poor day care."[46]

It was not just the specific tragedy of the Romanian babies that put Spitz, Bowlby, and Widdowson back on people's radar screens in the 1990s. This was also the decade that saw the rise of so-called third-wave feminism, with its attempt to reclaim more conventional gender roles as valid options for people. Third-wave feminists asked new questions about the effects of certain kinds of parental choices—to work outside the home, to put the kids in daycare—on children's well being. By now, Elsie Widdowson had become one of the most significant women scientists of her generation, developing path-breaking work in nutrition science. When she died in 2000, the obituary in the *New York Times* reviewed her other work, but saved her early orphanage studies—the best—for last. Why? In part, perhaps, because the perceived message of those early studies was now understood to function as a word from the wise to an anxious generation of *New York Times* readers. Her obituary writer concluded: "It turned out that love was the healthiest food of all."[47]

Is love good medicine?

And perhaps—just perhaps—love was even more than that. Perhaps it was also good medicine; perhaps healing ties could really heal, actually slowing or even reversing the course of an illness. As early as 1989, *Psychology Today* published an article called "Heart and Soul" that quoted the advice of one researcher to those prone to coronary heart disease: "Fall in love."[48] A decade later, in 1998, "heart health" expert Dean Ornish—

previously best known for his success in reversing risk factors for heart disease with a strict low-fat diet—published a new self-help book called *Love & Survival: The Scientific Basis for the Healing Power of Intimacy* that elevated that advice into an entire program. Diet was still important in controlling heart disease, Ornish wrote, but it couldn't compare with love: "Love and intimacy are among the most powerful factors in health and illness," he declared. "*I am not aware of any other factor in medicine—not diet, not smoking, not exercise, not stress, not genetics, not drugs, not surgery—that has a greater impact on our quality of life, incidence of illness, and premature death from all causes.*"[49]

This sounded like something a person would want to be true, but was it really the last word? In 2001, the National Heart, Blood, and Lung Institute published the results of a large multicenter clinical trial called Enhancing Recovery in Coronary Heart Disease (ENRICHD) that shook at least some people's confidence in healing ties. True, the goal of the study was not exactly to feed coronary patients generous helpings of love; but researchers did aim to use cognitive behavioral techniques both to improve levels of perceived social support in patients' lives and to reduce their levels of depression. For the 2,481 patients who participated, the behavioral techniques succeeded in both these aims. What they did not do, however, was reduce patients' risk of death or of a second heart attack. The quality of life of the patients in the active treatment group was better; but they did not live longer, on average, than the patients in the control group.[50]

Lisa Berkman, lead author of the 1976 Alameda County study, was one of the principal investigators on this study. In the fall of 2006, she spoke bravely and honestly to a class of my undergraduate students at Harvard about the disappointing results. "Have you ever seen 'nothing'?" she asked the students, and then presented a slide of a graph with the ENRICHD outcome data. "There: that's what 'nothing' looks like." At the same time, she said, she was not convinced that this was all there was to be said on the matter. Everything in her earlier professional life—from her time as a young woman working with uprooted and socially isolated people in San Francisco to her work on the Alameda County data—had convinced her that healing ties, on some level or other, were real. What felt wrong to her now was not the idea of healing ties so much as the belief that they could be demonstrated definitively by conducting controlled clinical trials on people who were already gravely ill. Perhaps what was needed instead, she suggested, was a more "public health" minded approach to the question, one that studied the effects of social connections in large populations across the life cycle.[51]

Another possibility, of course, is that healing ties are more effective for some disorders than they are for others. Maybe boosting social support cannot do much, if anything, to repair an already damaged heart; but maybe it can fortify an immune system and in this way enhance a person's ability to rid his or her body of, say, cancer. Indeed, the belief that cancer patients need community to heal—literally heal—was a claim that began to emerge strongly as early as the mid-1980s, against a backdrop of widespread feeling in the popular culture that the so-called war on cancer had largely failed.

As we have seen in earlier chapters, this feeling led in turn to a tendency to explore the holistic and alternative wings of medicine for the cures that were not forthcoming from orthodox medicine. In this context, some people became interested in the old psychoanalytic literature suggesting there might be a particular type of personality that was prone to develop cancer. The consensus over the profile of this personality was pretty clear: she (implicitly the stock image of this patient was always a woman) was emotionally repressed and, consequently, generally unable to experience intimacy or love on any authentic level. If the goal was now to help, rather than implicitly condemn, such a person, what might be done? The answer seemed clear: offer these patients opportunities to finally experience the intimacy, even love, that they had never previously risked or perhaps felt they deserved. Of those who spoke this way, none was more influential than cancer surgeon Bernie Siegel, author of the bestselling *Love, Medicine, and Miracles*:

> I feel that all disease is ultimately related to a lack of love, or to love that is only conditional, for the exhaustion and depression of the immune system thus created leads to physical vulnerability. I also feel that all healing is related to the ability to give and accept unconditional love. . . . The truth is: love heals.[52]

Highly popular with patients, this vision of love as a treatment for cancer—often conjoined with self-administered doses of positive thinking—was viewed with varying degrees of concern and outright anger by mainstream medicine. Most oncologists felt that cancer was a straightforward, serious biological disorder that needed to be fought with the best available biologically specific treatments. To their minds, it was a cruel deception to imply to patients that their minds had made them sick, and that they might therefore be healed by cultivating a different, perhaps more emotionally open and loving, approach to life.

Nevertheless, by the 1980s most mainstream cancer clinics did recognize the importance of offering patients opportunities for social support and counseling. These, it was acknowledged, were valuable ways of helping patients deal more effectively with the initial trauma of a cancer diagnosis, with the chronic anxiety, depression, and pain that typically persisted as treatment began, and with the possibility of a foreshortened future. In part, this new interest in social support reflected broader social changes in the 1970s, including the rise of consumerist approaches in medicine and the emergence of powerful patient-advocacy groups. It was time, some cancer patients began to say, to lift the long-standing taboo in American society against speaking openly about a diagnosis of cancer, to improve patient understanding of the disease and what could be done for it, and to agitate collectively for more research and better care. One catalytic moment here came when First Lady Betty Ford made the decision in 1975 to give an interview to *Time* magazine about her experiences of breast cancer. Another came a year later, when well-known NBC journalist Betty Rollin published a frank memoir of her own experiences with breast cancer, *First, You Cry* (later made into a television movie starring Mary Tyler Moore).

As the popular discourse about the emotional trauma of receiving a diagnosis of cancer became increasingly frank, the culture of care within oncology itself also began to change. A 1961 survey of physicians in a particular region of the United States found that more than 90 percent of them usually did not reveal a cancer diagnosis to the patient. A 1979 survey of physicians working in the same geographical region found, in contrast, that 97 percent of physicians now did so.[53] As these changes unfolded, a new field called psycho-oncology emerged, and began to insist that openness was not enough: patients needed to be helped to cope with the information, and to find ways to maintain what now came to be called "quality of life" throughout the difficult period of their treatment. As one of these new workers put the challenge in 1977: "As our colleagues in medicine explore new kingdoms of life-saving technology, we often find psychological wrecks in their wake. We can help to reconstitute these people."[54]

Although they were all familiar with the literature showing a relationship between social support and increased resilience to disease, the first generation of psycho-oncologists deliberately and clearly insisted that they were not engaged in a mind-body intervention; they were not trying to alter the course of disease but rather to improve quality of life. Indeed, many people in the field increasingly worried about the effects that popular ideas about the power of mind over cancer were having on patient morale. As Jimmie Holland, a psychiatrist at Memorial Sloan-Kettering Cancer Center in New York and one of the founders of psycho-oncology, put it in 1985, patients "feel guilty that they can't control their illness. . . . They look upon it as a weakness, partly as a result of the culture's emphasis on psychological means of controlling cancer."[55]

David Spiegel, M.D.
Courtesy of Dr. David Spiegel, Center for Stress and Health, Stanford University

Another person who felt like Holland was Stanford psychiatrist David Spiegel. Spiegel never forgot how a woman once telephoned him in a state of distraction and despair and explained how she had gone to a cancer support group to talk about her desperately ill son. The members of the group had looked at her and told her she needed first of all to realize that "every child with cancer is an unloved child."[56] Spiegel was outraged.

Like Holland, Spiegel believed

passionately in the importance of offering patients opportunities to improve their coping skills and enhance their quality of life as they learned to live with their cancer diagnosis. His own approach to that professional task had its origins in an experimental form of psychotherapy undertaken in the 1970s in collaboration with an older psychiatric colleague named Irvin Yalom. In a series of influential publications in the field of so-called existential psychotherapy, Yalom had argued that, in order to live authentically in this world, people must be willing to confront their deepest, most primal fear: not fear of their own sexuality, as Freud had thought, but fear of their own death.

> The primitive dread of death . . . resides in the unconscious—a dread that is part of the fabric of being, that is formed early in life before the development of precise, conceptual formulation, a dread that is chilling, uncanny, and inchoate, a dread that exists prior to and outside of language and image.[57]

In the 1960s, Yalom had thus pioneered forms of group therapy that had attempted to help patients confront their dread of death as a preamble to eventually learning to live fuller, more authentic lives. These were patients, however, who suffered from various emotional or neurotic disorders; none of them was actually facing the imminent possibility of death. In the course of a series of conversations, Spiegel and Yalom wondered how far a similar style of group therapy might help patients who were, in fact, terminally ill. Was it not possible that, by learning to look realistically and openly at their situations, and not hiding behind the various formulas of false optimism so popular at the time, they would in the end spend less energy defending themselves against their anxieties and more time appreciating whatever life remained for them? As Spiegel put it at the time:

> Professor Yalom and I believed that if these existential ideas are correct, they should work in action. People who are quite literally facing their deaths could actually live more fully and imbue their lives with vitality by reordering their priorities. But first, the considerable and deep anxiety that comes from staring into the bottomless pit of nonbeing had to be tolerated."[58]

In 1980, Spiegel and Yalom and another colleague undertook a study to see whether or not group psychotherapy treatment conducted according to Yalom's existential principles would in fact help people who were literally facing death. Eighty-six women with late-stage (metastatic) breast cancer—all receiving conventional medical treatments—were recruited. Fifty of these women were randomly selected to participate in weekly ninety-minute support group sessions, in which they were encouraged to talk openly about their fears and hopes and were taught simple self-hypnosis techniques designed to help them better manage pain and stress (Spiegel was also an accomplished clinical hypnotist). A year later, when the final data were gathered, it was found that the women who had participated in the group sessions did indeed have "significantly lower mood-disturbance . . . fewer maladaptive coping responses, and were less phobic" than the women who had only received conventional medical treatment for their cancer.[59]

It was a good outcome, and it could have been enough. It occurred to Spiegel, though, that he could use the data from this study to challenge

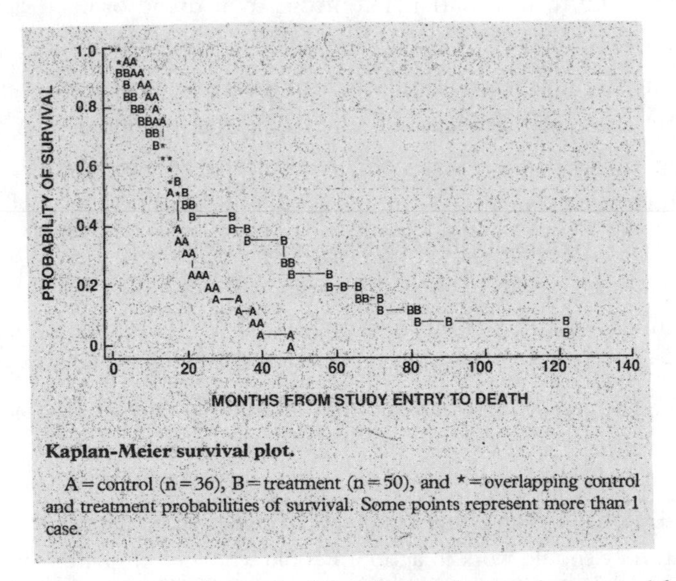

Kaplan-Meier survival plot.

A = control (n = 36), B = treatment (n = 50), and * = overlapping control and treatment probabilities of survival. Some points represent more than 1 case.

The results of David Spiegel's study of the effects of group therapy on life expectancy. Women who participated on a regular basis in weekly "supportive-expressive" group therapy sessions lived on average twice as long as controls. From *The Lancet* 2 (1989): 888–91. *Reprinted with permission from Elsevier* the "false hopes" peddled by what he called the "wish-away-your-cancer crowd."[60] And this was when he got an unexpected surprise. For when he reviewed his data, Spiegel found that the support groups appeared to be doing more than just improving the mood and coping of the women in the active treatment group. They also seemed to be helping these women to live longer—on average, twice as long (36.6 months as opposed to 18.9 months) as the women in the control group.[61] Only three women from the original study were still alive at the point that Spiegel was examining his data, and all three had participated in the support groups.

It was the first time a randomized clinical trial—the gold standard of medical research—had ever been used to test the power of healing ties, and people were floored. Spiegel's findings were published in *The Lancet* alongside an editorial that urged others to find an example in "this intellectually honest approach" and noted that its methodology seemed to be "beyond criticism."[62] For cancer patients, the name of David Spiegel suddenly became, in the words of one commentator, "one to conjure with."[63] Press coverage at the time helped to encourage this: "This is a marvelous study, a surprising study to me as well," one of Spiegel's colleagues told the *Los Angeles Times*. "I would have bet the mortgage of my house that it would not have come out this way."[64] Before long, Spiegel and some of his patients had accepted an invitation to appear on the television talk show *Oprah*, and a few years later his group therapy was profiled in the Public Broadcasting Service's 1993 blockbuster documentary *Healing and the Mind*. There on television, before millions of viewers, journalist Bill Moyers asked Spiegel, "If the findings of your study are replicated, what do you think it means for medicine?" Spiegel responded forcefully:

> [W]hat it means is that we have to change the definition of what health care is. . . . It means that health care is more than just physical intervention. It's [also] support from a caring physician and health care team and some kind of group intervention to help people who are seriously ill learn how to cope with it as fully as possible. That would be a wonderful change in the direction of health care and a cost-effective addition to helping people live better and perhaps live longer.[65]

By the mid-1990s, the word was out on the street that, to maximize one's chance of recovery from cancer, a person should be prepared to undergo not just sessions of chemotherapy or radiation but also regular sessions of bonding with other cancer patients. Here is how one patient, looking back in 1999, remembered that time:

> When I was first diagnosed with breast cancer five years ago, a number of well-meaning friends and family members suggested that I join a support group. "You'll live longer," they said, citing articles they had read in their local newspapers linking support group participation to improved health. Many of these news reports mentioned Stanford psychiatrist David Spiegel's 1989 landmark study suggesting that support group participation may have increased survival time for women with metastatic breast cancer. . . . While I did not have metastatic breast cancer, I quickly added support group participation to my cancer survival checklist—giving little thought to whether it would feel good. . . . What I really wanted to do on the cold Boston winter day that the support group met, was to sit home and watch "Oprah" while drinking some Starbucks coffee.[66]

For Spiegel himself, all these developments meant something different: that he was suddenly forced to keep company with alternative practitioners whose approaches he had spent years reviling. He found it an uncomfortable experience, and went on to write his own self-help book designed to correct the excesses of theirs. This move led one reviewer to grumble that "Spiegel is too dismissive of some workers in the field . . . who had trodden the same path before him. It ill becomes him, as a late convert, to 'knock' other doctors who have undoubtedly helped and inspired many people with cancer. There is room for them all."[67]

In fact, of course, Spiegel had not trodden quite the same path in his career as these other workers. His had been a journey from existentialist psychiatry, that is, from a form of clinical practice that had actually been designed to help women confront their deaths, not extend their lives. By design, the group therapy sessions Spiegel conducted possessed a starkness and (frequently) a darkness that were barely compatible with the usual understandings of what group therapy for cancer was supposed to be good for.

> The support groups we formed offered far more than hand-holding and good wishes. The sessions became a time and a place for women to express some of their deepest fears, what one called "that sense of waking up at three in the morning with an elephant sitting on your chest." She added: "I wonder if I will live to see my son graduate from high school or my daughter get married. I have to keep up a front everywhere else—that's so hard."[68]

Then things began to unravel. In 2001, a multisite study failed to replicate Spiegel's findings. There was no evidence, its authors declared, that participation in group therapy extended life in women with metastatic breast cancer (the study did confirm that participation in such groups improved mood, experience of pain, and coping skills).[69] The Associated Press was quick to spread the news in a report taken up by multiple newspapers and trade magazines: "Therapy doesn't extend breast cancer patients' lives."[70]

Spiegel's own response was to insist that the last word had not yet been said on this matter. There might be at least two reasons, he said, why the replication had failed. Improvements in conventional cancer treatment since the 1980s, for example, might be masking the independent impact that group therapy really does have on the course of disease. In addition, since most educated cancer patients by now had heard that joining a support group was likely to improve their odds of survival, it was likely that, when they were randomized to the control group, some patients looked outside the study for other forms of social support and group therapy.[71] Spiegel himself, who had been due to complete a replication study of his own in 2000, decided he was not ready to publish. His own data were not clearly supporting the original hypothesis. His team instead sought funding for an additional five years.

As of this writing (March 2007), Spiegel remains unwilling to say that support therapy does not extend the life of women with cancer; he and his team believe there is still some kind of biological story to be told about the power of healing ties in the case of cancer, even if it might not be quite the story with which they started.[72] At the same time, in recent years Spiegel has also increasingly returned to the original existential rationale for the kind of psychotherapy he offers patients: to help them live more authentically and fully, no matter how long or short the time they have remaining. "It's clear by now that this kind of psychotherapy helps cancer patients," he says. "How it helps them is a matter of legitimate disagreement. That it helps them is not."[73]

In fact, many of the women who participated in Spiegel's replication study would likely agree with him. Some years ago, during the time when interest in Spiegel's original findings of enhanced survival was at its peak, he gave me permission to go to California to talk with some of these women. I was interested above all in asking them a simple question that it seemed no one had ever asked: namely, what did they think? Living this particular "healing ties" story from the inside, did they believe it? Did they think it was helping them to live longer?

I asked the question, and I will never forget what happened. A silent snort went around the table. No, they said, they did not believe the premise of the study—not really. Why not? I asked. Their answer was clear: the evidence was not there for them; they had seen too many people in their group die. Then they said something else that surprised me: from their point of view, it did not really matter if the replication study failed. "It would be nice if David's hypothesis proves out and maybe it will in the long run," one said, "but I don't think it matters to me at all. That's not why I joined the group."

Why, then, were they in the group? I asked. What did it mean to them? Again, the answer was clear: they stayed in the group because they learned there, from one another, how to live better with cancer and how to die better from cancer, something that they could learn nowhere else in their culture. And more particularly, they had all learned that the process of dying was infinitely eased when one did not die alone; as a group, they had learned that they could give this gift of connection and companionship to one another.

At this point in the conversation, a patient who had been largely quiet so far looked up and perhaps took pity on me. She knew I had come to ask whether they believed that participating in the group was helping them to live longer, and perhaps she thought I was disappointed by their response. So she tried to help me by putting the consensus of the group in a different way. "If you eliminate the concept of time," she told me, "I guess then that you could say that we live longer."[74]

Chapter Six

EASTWARD JOURNEYS

Scientists study it. Doctors recommend it. Millions of Americans—many of whom don't even own crystals—practice it every day. Why? Because meditation works.

—*"Just Say Om,"* Time *magazine, August 4, 2003*

Don't Let Ancient Knowledge Become a Thing of the Past . . .

—*Cover slogan of the magazine* Qi: The Journal of Traditional Eastern Health & Fitness

On February 4, 2003, the *New York Times* published an essay by Daniel Goleman, a well-known science journalist and psychologist with a long-standing interest in meditation. The piece was titled "Finding Happiness: Cajole Your Brain to Lean to the Left," and in it Goleman hailed what he called a growing "rapprochement between modern science and ancient wisdom." He spoke of how, as a graduate student, he had wondered whether traditional Asian meditation practices "might work as an antidote to stress" but had been discouraged from extensively pursuing this idea by the skepticism of his professors and the inadequacy of the laboratory measures available to him.

Now, however, he said he felt "vindicated." Science today stood at the brink of understanding "the brain mechanism that may account for meditation's singular ability to soothe." More particularly, new work studying the effects of meditation on the brains of Tibetan Buddhist monks suggested that ancient meditative techniques might result in neurological changes conducive to long-lasting happiness and well-being.

In 2000, University of Wisconsin neuropsychologist Richard Davidson had traveled to Dharamsala, India, the seat of government for the exiled Dalai Lama of Tibet. There, surrounded by both other Western scientists and Buddhist monks, he shared the relevant research with the Dalai Lama. The spiritual leader was intrigued. In Goleman's words, "was there something about the training of lamas—the Tibetan Buddhist equivalent of a priest or spiritual teacher—that might nudge a set point into the range for perpetual happiness? And if so, the Dalai Lama wondered, can it be taken out of the religious context to be shared for the benefit of all?"[1] Further scientific research, Goleman said, might soon help us find out.

In one sense, the story that Goleman told was all about new and specific things: new scientific findings, a new conclusion about the potential universal benefits of meditation, a specific encounter between a scientist from the West and a spiritual leader from the East. Nevertheless, the article was effective as a piece of journalism in part because the story it told was, on a structural level, neither new nor specific at all. It drew, rather, on the tropes, values, and literary sensibilities of a particular narrative of mind-body medicine that I call "Eastward journeys."

What is this narrative? It begins, like "healing ties," as a response to the mood of broad antimodernist lament that has characterized many American stories about stress since at least the discontented 1970s. The stress of our modern way of life in the West, it says, has damaged our hearts, undermined our immune systems, and made us far more unhappy than we had ever imagined we would be. Nevertheless, there is hope for us. Though we may not be able to set right what ails us, let alone fix all that has gone wrong in our society, there are others who can.

Who are these others? They come, says this narrative, from the East, and they are everything that we who live in the West are not. We are modern; they have stayed in touch with ancient traditions. We are harried and tense with stress; they speak and act wisely from a place of deliberate repose and contemplation. Our medicine is aggressive and treats the body as if it were a broken machine; theirs is gentle and guided by an appreciation of the suffering of the whole person. We are doggedly dualistic in our ways of thinking about the mind-body relationship; they understand the extent to which mind and body are intimately intertwined and have found ways to translate that knowledge into uniquely effective healing practices. The solution to our woes, then, is simple: we must find ways to learn from these people of the East, and in that process of learning, discover ways to heal ourselves.

How is this to be done? The concept of journey—and encounter—moves the narrative to its conclusion. East must meet West, and knowledge must be shared. The encounter may happen in many different ways. Western scientists may travel to Dharamsala and sit in the private quarters of the Dalai Lama, surrounded by monks in saffron robes and other exotic reminders that one is far from home. Western television journalists may travel to China to sniff strange-smelling concoctions and bear witness to exotic methods of mind-body healing. Buddhist monks from Nepal may travel to the United States to meditate inside fMRI machines. The Dalai Lama himself may travel to Harvard University or MIT or the University of Wisconsin to discuss meditation with brain scientists. At a different level, ordinary people may embark on private Eastward journeys of their own, buying books full of Asian spiritual teachings, venturing into unfamiliar clinics and centers, or apprenticing themselves to teachers who train them in exotic practices like qigong. Regardless of the way it happens, the moral of every successful "Eastward journeys" story is the same: we can have the best of the East without abandoning all that we value from our own traditions.

While the immediate origins of the "Eastward journeys" narrative lie in the counterculture movements of the 1960s and 1970s, the habits of thinking on which this narrative relies go back as far as the first sixteenth-century European ventures into Asia for the purposes of trade, missionary work, exploration, and eventually colonization. Over the course of several centuries, there arose a convention of writing about the native people of those countries that was both patronizing and exoticizing. The accounts of Christian missionaries, travelers, and colonial administrators were filled with descriptions of primitive despots, self-flagellating penitents, bizarre yogis, and indolent opium eaters. Such descriptions were then generally contrasted, implicitly or explicitly, with the civilized, rational, and pragmatic behaviors of the West. This convention of turning Asian people into the inverted mirror image of their Western counterparts was christened "Orientalism" in 1978 by the Palestinian literary critic Edward Said. Said himself was particularly concerned with the role this convention had played historically in advancing European colonialist and imperialist agendas.[2]

Beginning in the early nineteenth-century, Orientalist conventions of writing and thinking were also used in the United States and Europe by critics and radicals who effectively reversed the original moral logic of this tradition. Still stylized, still exoticizing, this new, more romantic form of Orientalism used idealized images of the East to highlight Western moral and spiritual failings. Thus, in the early nineteenth century, transcendentalist philosophers like Ralph Waldo Emerson and Henry David Thoreau celebrated the compatibility of ideas they read in the newly translated Bhagavad Gita with radical ideals to which they themselves were committed—individualism, self-reliance, and the priority of experience over faith.[3] In the last decades of the nineteenth century, the Russian founder of Theosophy, Helena Petrovna Blavatsky (later naturalized as an American) and her supporters spread the doctrine that the literature of the East possessed stores of ancient wisdom that the West had once also possessed but had long since lost or forgotten.[4]

By the second half of the twentieth century, as colonialism became a shameful legacy and Western cultures continued to grow more ambivalent about their modern values and lifestyles, the romantic variant of Orientalism—an Orientalism dominated by visions of ancient teachers, texts filled with occult secrets, meditating monks on misty mountaintops, and serene sanctuaries—gained a new lease on life, especially within alternative, countercultural circles.[5] "Eastward journeys," focused on mind-body healing, draws on the romantic and self-critical tropes of this kind of Orientalism, and then adds one more of its own. It says the East is not only a spiritually but also a medically exemplary place. Its traditional doctors and spiritual teachers are skilled in ways of mind-body and holistic healing that we have lost, forgotten, or simply never known. The fact that many of the practices in question are supposed to be particularly effective for stress—a state Westerners associate with all the excesses and distorted priorities of their modern lifestyles—is telling. When we tell stories about "Eastward journeys," we seem to be saying that what modernity has wrought, ancient wisdom will heal.

From the Maharishi to medicine: meditation as a stress buster

It is one of the iconic images of the late 1960s: the Beatles at an ashram in India wearing necklaces of beads and flowers and posing with their shaggy-headed, affable teacher of meditation, the Maharishi Mahesh Yogi.[6] Clearly, they were having an Orientalist encounter, but they were not on an Eastward journey in the sense that I am using the term. That is to say, there was no therapeutic motivation behind their desire to study meditation with the Maharishi, no hope that meditating might reduce their stress levels or make them healthier. What then was their motivation? As Paul McCartney much later explained (somewhat incoherently) in an interview that was broadcast on television in 1995, "We'd been into drugs, and we were—there's the next step, then, is—then you've got to try and find a meaning, then." George Harrison, also participating in this interview, then clarified, "That's where [we] really went for the meditation."[7]

The Beatles were not alone in thinking of meditation as the next thing to try after psychedelic drugs. As one young woman in New York City explained to a journalist in 1967, "I kept thinking that through the constant use of LSD, I'd return to the religious feeling I had with it the first time. . . . But it never came and I met Swami. I gave up drugs. I was hooked on the religion and on yoga. I'm a better person now. I'm not hung up on myself anymore." She may have thought this, but a teacher of Hinduism living in the United States grumbled to that same journalist that, in his view, young people were coming to the ashram simply to get a drugfree ride to ecstatic experiences: "They are exhibitionists. They have no discipline and what are they really learning about Hinduism? This trend toward a drug culture is very dangerous."[8]

Dangerous or not, the blending of psychedelics with Asian mysticism was all part of the more general 1960s and 1970s countercultural fascination with Hinduism, Buddhism, and more generally, all things Eastern. Asian spiritual traditions were believed to be fundamentally experiential and mystical and therefore to possess an authenticity and depth that homegrown Judeo-Christian alternatives lacked. In the words of one popular teacher of Zen Buddhism at the time, Alan Watts: "Obviously, if

Christian groups cannot or will not provide mystical religion, the work will be done by Hindus, Buddhists, Sufis, unaffiliated gurus, and growth centers. Church men can no longer afford to laugh these things off as cultish vagaries for goofy and esoteric minorities."[9]

Some people learned about Asian spiritual practices and philosophies from charismatic Asian teachers and writers who had chosen to reach out to Westerners: Paramahansa Yogananda (author of the perennial bestseller *Autobiography of a Yogi*), Chögyam Trungpa (a high-energy Tibetan Buddhist lama who founded the first accredited Buddhist center of higher education, Naropa University, in the United States), Thich Nhat Hahn (a Vietnamese Buddhist teacher who integrated his teachings of contemplative practice with a strong call for social justice), Krishnamurti (a radically nonsectarian teacher from India, who rejected the efforts by Theosophists to embrace him as a "World Teacher"), and so on.

Others learned from Western writers who had tasted the Asian alternative and aimed to share the results with their readers. Particularly important here were the novels and poetry of the 1950s Beat writers: Gary Snyder, Jack Kerouac (author of the thinly disguised memoir *The Dharma Bums*), and Allen Ginsberg.[10] More didactic books like Alan Watts's *The Way of Zen* (1957) and *Psychotherapy East and West* (1961) provided further routes into the world of Asian wisdom.

Where did the Maharishi Mahesh Yogi fit into this larger scene? Born Mahesh Prasad Varma in 1917, the man who would one day be dubbed the "Chief Guru of the Western World"[11] was raised in a comfortably middle-class family in India.[12] His father had been a minor official in the Department of Forestry, and the Maharishi himself was given a secular, Western education, up to and including university-level study of physics and mathematics at Allahabad University.[13] By his own account, this upbringing only had the effect of creating within him a deep spiritual restlessness, and a resolve to resist the conventional path his parents had intended for him. In consequence, he embarked on his own journey in search of Eastern wisdom, apprenticing himself for thirteen years to a Hindu swami, Brahmananda Saraswati (later known to Westerns as Guru Dev). Brahmananda Saraswati was revered by his followers for many things, but none more than his deep knowledge of the ancient Sanskrit texts the Vedas.

Among other things, the Vedas teach meditative techniques organized around the repetition of mantras that are supposed to facilitate identification with the energies of a range of deities. Mahesh Prasad Varma became convinced that, properly modified, these ancient techniques might be of great value to the modern world. He decided it was his destiny to find a way to take them out of relative obscurity and teach them to those who would otherwise never be exposed to them. Taking the honorific title Maharishi ("great sage") in 1955, he began his teaching in India. But from the outset, he had his eye on winning over the West. His first trip to the United States came in 1959.

Initially the Maharishi was just one of a crowd of gurus and teachers; and the crowds that found the time to come hear him numbered in the hundreds rather than the thousands. Even so, his goals were ambitious: nothing less than the spiritual regeneration of the world through the teaching of a simple meditative technique he had distilled from ancient Hindu practices. "These are ancient teachings," the Maharishi later explained, "but now I have made them new."[14] He called the practice transcendental meditation, or TM.

The secret of the practice, he taught, lay in the mantra—the special word or phrase students learned to repeat during their practice. Each mantra was supposedly tailored to the unique spiritual needs of the student, and each was said to possess a resonant spiritual energy capable of "draw[ing] attention of those higher beings or gods living [in some other world]."[15] Students were asked to pay a rather stiff fee for their mantra, which was imparted to them with great ritual, secrecy, and reverence. (Today, it is known that all the mantras were actually chosen from a list of seventeen Hindu deities and given according to the age of the student.)[16] The Maharishi taught that by spending just fifteen to twenty minutes twice a day in meditation, repeating their chosen mantra, students could hope to experience a special state of awareness called "pure consciousness" that would, over time, make them happier, more peaceful, and more intelligent.

All of this might have remained a minor part of the larger marketplace of Eastern practices and teachings on offer in the 1960s, had it not been for the fact that, encouraged by George Harrison, the Beatles decided that the Maharishi should be the one to guide them in the next phase of their spiritual journey. Once they made that decision, other celebrities followed their example: football star Joe Namath became a follower, as did movie stars Jane Fonda and Mia Farrow.

Suddenly the staff who worked to promote the Maharishi's teachings could hardly keep up with demand. Before the end of 1966, records kept by the staff show that only about a thousand people had learned TM, but from 1966 until 1975 the practice experienced rapid growth, culminating with the initiation of 292,517 people in that single year.[17] The Maharishi himself was invited to appear on American television talk shows and gave innumerable interviews. The famous *New York Times* piece proclaiming him to be the "Guru of the Western World" was published in 1967.[18] All and all, the Beatles were very good news for the Maharishi and his movement.

In 1968, the Beatles, accompanied by wives and friends, made a widely publicized journey to India to participate in a two-month meditation course with the Maharishi at his ashram in Rishikesh, India. Ringo Starr and his wife went home after just ten days, complaining about insects and the food, but the rest of the group stayed on. (Most of the songs for the *White Album* were composed at the ashram in the evenings during this period.)

Then things began to unravel. The Beatles became persuaded (probably wrongly) that their teacher had made a series of sexual advances on one or more of the women in the entourage, including Mia Farrow. A disillusioned Paul McCartney left first, followed a few weeks later by George Harrison and John Lennon. When the bewildered Maharishi begged for an explanation, Lennon replied curtly that if he really was "so cosmic" then he'd "know why." In the airport, waiting for his plane, Lennon wrote the song "Sexy Sadie," with its opening line, "Sexy Sadie, what have you done? You made a fool of everyone."[19] Later, in an interview, Lennon admitted that he "copped out and wouldn't write 'Maharishi, what have you done, you made a fool of everyone.' " He then leaned into the mike of the tape recorder: "But now it can be told, fab listeners."[20] The romance between the Maharishi and the Beatles was over.

And so too, in a sense, was the romance between TM and the counterculture. Within the upper echelons of the Maharishi's organization, a consensus was growing that the TM movement needed to distance itself from the counterculture and its fickle celebrities. It is true that the Maharishi did embark on a tour (by all accounts, ill advised) with the Beach Boys later in 1968. He also accepted an invitation to speak to the thousands of young people attending the Woodstock concert in 1969, where he urged them to help bring about world peace by transforming their consciousness. Nevertheless, it was clear that his reputation was not what it had been, and that the marketing of TM would need to proceed on a new basis. This new basis was found in the decision to emphasize the potential compatibility of TM with scientific thinking and practices, to attempt to make inroads with scientists in universities rather than with pop stars.

The new outreach effort worked: conferences were organized, and scientists came. These were years when it was fashionable to argue for deep structural relationships between the claims of Eastern mysticism and the claims of quantum physics, and there was some excitement in the early meetings about the possibility that the mind-altering effects of TM might be explicable as a "quantum event."[21] Perhaps, some early participants suggested, it could be defined as a "super-conductivity effect" (a quantum mechanical phenomenon), or as a "macroscopic quantum state in the brain."[22]

Then, in 1969, a graduate student at the University of California, Los Angeles decided to research the physiological effects of TM for his dissertation, and almost single-handedly changed the focus of the scientific conversation about TM. Robert Keith Wallace recruited college students who had taken a course in TM. He hooked them up to various measuring instruments, asked them to meditate, and found that on average they showed significant changes in their physiological state: reductions in oxygen consumption, reductions in resting heart rate, and changes in skin resistance.

But that was not all. Wallace also carried out electroencephalographic (EEG) recordings of his students while they meditated, and he found that TM practice was associated with a highly coherent pattern of brain wave activity, one he believed to be different from anything previously reported in the literature. The Maharishi and his followers had long claimed that TM practice produced a unique state of consciousness. Wallace, it seemed, had now proven them right.

In 1970, Wallace announced his discovery of a "fourth major state of consciousness" in the flagship journal *Science*:

> Physiologically, the state produced by transcendental meditation seems to be distinct from commonly encountered states of consciousness, such as wakefulness, sleep, and dreaming, and from altered states of consciousness, such as hypnosis and autosuggestion.[23]

Wallace's principal focus in this article was the phenomenon of altered states of consciousness. At one point, he did suggest that some of the global physiological effects associated with TM—the reduction in oxygen consumption, the changes in resting heart rate, and so on—might have clinical implications; but these were clearly of less immediate interest to him than the EEG findings. Harvard Medical School cardiologist Herbert Benson, however, took a different view. In the late 1960s, Benson was one of the small but growing number of cardiologists who had come to believe that stress played an important part in heart disease. As he explained to a journalist in 1975, "many of the problems that heart doctors encounter [such as dangerously elevated blood pressure] have been created by daily stresses and tensions—the cost, so to speak, of living at an often hectic pace in a highly complex society."[24]

Could anything be done about this? Benson tried to find out. In one study, he trained squirrel monkeys to raise their blood pressure using operant conditioning technology. As he recalled later, "We found that the monkeys that were 'rewarded' for higher blood pressure went on to develop hypertension.[25]

If monkeys could learn to raise their blood pressure through operant conditioning, was it also possible to teach them to lower it and, in this way, create a behavioral therapy for hypertension? That was the question Benson began to pursue. At the same time, he and some colleagues began to experiment with a clinical intervention for human subjects that would draw on similar operant conditioning techniques. They used biofeedback to alert subjects to the state of their blood pressure levels, and all changes in a downward direction were reinforced with a flashing light, a sound, and attractive slides flashed on the screen (including reminders of the amount of money the subjects were earning for participating in the study). In the end, six out of seven subjects who went through this proto-

Herbert Benson, M.D., photograph taken in the summer of 1972, when he was first actively developing the concept of the Relaxation Response.
Courtesy of the Harvard Medical Library in the Francis A. Countway Library of Medicine

col learned to lower their blood pressure levels, at least in the experimental setting. The results of this work were published in *Science*.[26]

It was a busy time. Then a chance encounter changed everything. Benson was approached by a group of TM practitioners who told him he should stop studying monkeys and start studying them. They did not need to rely on cumbersome conditioning techniques, they told him; the daily practice of TM allowed them to lower their blood pressure at will. At first, Benson refused; meditation was a fringe counterculture practice without any perceived medical implications, and he could see no reason to shift the focus of his research in this direction. But the young TM practitioners persisted, and finally Benson relented. He agreed to look into their claims, but on the condition that all the studies were done in the evening, and that the students came to the laboratory through a back door so that none of his colleagues would see them.

Leery of how this new work would be received, Benson tested the waters with a letter to the editor of the *New England Journal of Medicine* in 1969. The letter said nothing about the physiological research he was undertaking with TM practitioners, but instead focused on an issue of obvious interest to the medical community at the time: drug abuse. He had learned from his subjects, he told the readers of the *NEJM*, that, prior to beginning their practice of TM, many had indulged in recreational drug use: marijuana, LSD, and in several cases, heroin. Now, however, all of them "reported that they no longer took those drugs because drug-induced feelings had become exceedingly distasteful as compared to those experiences during the practice of transcendental meditation."[27]

Here, then, was a first cautious suggestion that TM might have serious medical benefits.

Bolder claims were still to come. When he first began studying the TM practitioners, Benson had not known of Robert Keith Wallace's work; but on discovering it, he proposed a collaboration. Wallace moved to Harvard, and he, Benson, and a third colleague, Archie F. Wilson, developed a new protocol to study their subjects. Blood pressure, heart rate, brain waves, rates of metabolism, and rates of breathing were all to be measured under three conditions: first, the subjects would be asked to sit quietly for twenty minutes; and second, they would be asked to sit quietly and meditate—repeat their mantra, etc.—for twenty minutes. They would then be asked to return, for a final twenty minutes, to a state of quiet repose. The aim was to assess the distinctive contribution, if any, of meditation. "What we found," Benson later recalled, "was astounding. Through the simple act of changing their thought patterns, the subjects experienced decreases in their metabolism, breathing rate and brain wave frequency."[28]

It wasn't the altered states of consciousness observed in his meditating subjects that astounded Benson—so far as he was concerned, the patterns of brain wave activity seen in their EEGs were evidence merely that they were very relaxed. What astounded him, rather, were the effects that meditation produced on visceral and autonomic functioning. Taken together, these seemed to amount to a systematic reversal of the fight or flight, or stress, response. Later, Benson liked to talk about the fact that, by serendipity, the discovery of this reversal effect had been made in the very same laboratory where, two generations earlier, Walter B. Cannon had discovered the fight or flight response itself.[29]

It is important to notice here the ways in which, in doing this work, Benson both did and did not endorse the Maharishi's central claims about TM. On the one hand, he found clear evidence for widespread physiological effects of meditation on both brain and body; on the other hand, he did not regard these effects as having anything to do with the specific mantras that were silently repeated by his student subjects, and he certainly didn't regard them as being in any way connected to deities or unseen energies. From Benson's point of view, the testing suggested that any word or phrase, even a meaningless one, repeated for a sufficiently long period under quiet conditions was capable of producing the same relaxed state.

Benson had previously been lionized by the TM community and personally applauded by the Maharishi, but his position on the mantras now made him persona non grata in those circles. ("We got into an enormous argument with the TM people," he ruefully recalled many years later.)[30] Wallace and Benson severed ties over this issue (Wallace went on to become the first president of the newly founded Maharishi International University), and Benson continued his work with other colleagues. By 1972 he was clear: whatever other functions they might have had, TM and all the other ancient meditative techniques central to faith traditions around the world also acted as technologies for turning off the stress response. Ironically enough, this made them of greater importance than ever to individuals living in the modern age. As Benson told a journalist at this time, "In modern society this fight or flight response is often an anachronism, and such stimulation of the sympathetic nervous system may lead to diseases like hypertension. TM appears at the present time to be the easiest and most rapid way of turning on an opposite response."[31]

Benson was personally persuaded of the importance of TM, but he remained nervous about the wider implications of what he was doing. For one thing, there was his professional position to think about: for a Harvard professor to commit himself to studying the health benefits of meditation struck many of Benson's colleagues as professional suicide.[32] For another, there was the question of the religious implications of his work. "After seeing . . . how much this was [practiced] in a religious context, it hit me, this is prayer, this is one form of prayer. And I got frightened, really scared." Benson went to see the dean of the Harvard Divinity School and asked for his advice.

> I told him my worry about what I was doing and the potential that I might be undermining religion. Stanhope [the dean] stood up, came over to me, leaned down, a towering, very thin sort of figure and said, "Young man, religion was here before you and religion will be here after you. You do your thing, we'll do ours. Good-bye." That was it. So I figured, "the hell with it, I'll do my thing."[33]

Meditation for a counterculture generation. *Time* magazine cover from October 13, 1975. *© Time Life Pictures / Getty Images*

In 1974, Benson formally eliminated any remaining associations between his research and TM by renaming the effects he was studying "the relaxation response."[34] And a year later, he published his book *The Relaxation Response*.[35] Ironically, the book was published in the same year that *Time* magazine published a feature article on meditation beneath a lurid cover image of the Maharishi Mahesh Yogi painted in brilliant colors and set against a backdrop of wildflowers.[36] Clearly, Benson was going to face an uphill battle in his efforts to distance meditation from the Maharishi and win it a hearing within the medical mainstream.

He had less trouble convincing the vast population of patients and potential patients in the popular culture to listen to him. The year it was published, *The Relaxation Response* went to number one on the *New York Times* best-seller list; today it is in its sixty-fourth printing, has been translated into fourteen languages, and has sold some five million copies. Cowritten with a professional science writer, it used a number of distinctive rhetorical techniques to try to persuade readers to see the relaxation response Benson's way. It offered a first-person account of Benson's scientific odyssey toward the discovery of the response; it represented the response as an emergent product not of the counterculture but of the world of stress research, clinical studies of heart disease, and modern medicine; it provided a plain-talking discussion of the physiological evidence for the existence of the response; and it gave readers a step-by-step guide to the practice of the response.

Finally, the book went to great pains to dissociate meditation from its Eastern associations. Methods for evoking the relaxation response, Benson observed, existed in virtually every known religious tradition, without being the sole possession of any of them. In other words, one did not need to be Buddhist or Hindu to meditate. Indeed, one did not even need to be religious, because the techniques worked regardless of faith or belief system. "Even though it [the relaxation response] has been evoked in the religions of both East and West for most of recorded history," Benson reassured his readers, "you don't have to engage in any rites or esoteric practices to bring it forth."[37] In short, Benson domesticated and medicalized meditation. As one journalist put it in 1975, Benson had helped her appreciate meditation as "a terrific aspirin, a wonderful kind of bromide."[38]

The endorsements printed on the frontispiece of the original edition of *The Relaxation Response* were carefully calculated to reinforce this resolutely secular and pragmatic message. No religious leaders or meditation teachers were asked to comment. Instead, the book sported puffs from the executive editor of the *Harvard Business Review*, a cardiologist, a general practitioner, and a leading stress researcher. "I am delighted that someone has finally taken the nonsense out of meditation," said one of these men. "This is a book that any rational person—whether a product of Eastern or Western culture—can wholeheartedly accept."[39]

By the 1980s, this new medicalized understanding of meditation was no longer strictly identified with Benson's secularized version of TM. It was also now being extended to other kinds of meditation, especially so-called mindfulness or vipassana meditation from the Buddhist tradition. Instead of repeating a mantra over and over, students learning this tradition practiced stabilizing their attention in ways that allowed them to introspect on their experience without reacting or judging. Over time, it was believed that such efforts enhanced self-understanding, equanimity, clarity of mind, and compassion. Some American teachers of Buddhism described a mind possessed of such qualities as a "healthy mind," but it is important to realize that this term was a translation from the Pali word *kusala*, which refers to qualities of merit or virtue as much as health, or rather makes no distinction between these concepts. It was not a medical claim.[40]

To say this, however, did not mean that the cultivation of such "mindful" virtues might not be valuable to people who were challenged by chronic or serious medical disorders—or at least that was the conclusion of a young American teacher of Buddhist meditation and yoga (who also happened to hold a Ph.D. from MIT in molecular biology) named Jon Kabat-Zinn. In 1979, he persuaded officials at the University of Massachusetts Medical Center in Worcester to let him set up a program in which patients suffering from chronic pain or other chronic disorders would be trained in "the regular, disciplined practice of moment-to-moment awareness or *mindfulness*, the complete 'owning' of each moment of your experience, good, bad, or ugly."[41] In time, he believed, patients would learn that a considerable amount of their suffering had to do less with their pain itself, and more with their emotional reactions to that pain.

Kabat-Zinn called his program a course in "stress reduction," but not because he believed that mindfulness meditation was a device for turning off the stress response in the way Benson had claimed TM did. In fact, Kabat-Zinn and his staff liked to joke about how "stressful" it could be to go through their demanding eight-week course. His decision to use the language of stress reduction to describe his work at the medical center was actually more strategic than conceptual: speaking of stress reduction rather than meditation was, he felt, a nonthreatening way to persuade both the medical establishment and ordinary patients to test the potential of a practice they might otherwise have dismissed as kooky or cultish: "If you go in talking about the Buddha and inviting masters with shaved heads for lectures," he told an interviewer in 1991, "it's going to be perceived right away as some foreign cultural ideology—a belief system." He went on to insist to this same interviewer: "The clients are not being sent to learn to heal; they're not being sent to have their symptoms go away or to master them. They're being sent as a palliative, to help them become more calm."[42]

Nevertheless, over time, Kabat-Zinn became increasingly disinclined to draw sharp distinctions between the medical and the existential value of this practice. It turned out that patients who took his program were not only less anxious and depressed but also frequently reported less physical pain.[43] Later, he was able to secure evidence that a simple mindfulness-based practice might accelerate the treatment of psoriasis using phototherapy,[44] or even fortify immune system functioning while shifting the "emotional set-point" in the frontal cortex.[45] Could it be that the cultivation of Buddhist mental virtues led over time to a physically healthier body and even brain?

By the 1980s, programs modeled on Kabat-Zinn's approach, now known as MBSR (mindfulness-based stress reduction), were being set up in other clinics and hospitals. Kabat-Zinn's 1990 introduction to his method in *Full Catastrophe Living: Using the Wisdom of Your Body and Mind to Face Stress, Pain, and Illness* has now been in print through twenty-six printings and a fifteenth-anniversary edition was released in 2005.[46] Throughout these developments, Kabat-Zinn remained committed to his original strategy: downplay the Eastern exoticizing side of these techniques and invite patients instead to discover their practical value for themselves through personal experience. In a conversation with me in early 2007, he mused that:

> Alan Watts would be amazed, as would be Carl Jung, at the degree to which Buddhist meditative practices have penetrated Western

clinical medicine and psychology in this era, not as some kind of skin graft, but as a deeply personal and serious scientific exploration that has already shown significant benefits to at least tens of thousands of people, but that has little to do with a Journey to the East.[47]

From Mao to Moyers: Qi, China, and the invention of an ancient tradition

Kabat-Zinn's perspective surely captures a partial historical truth, but there is more to say. Even as efforts were made on one front to thoroughly secularize and disentangle one mind-body practice from its Eastern associations, the public was being offered a different, thoroughly exotic new vision of mind-body healing from the East: one grounded in previously little known principles of traditional Chinese medicine.

A critical catalyst here came in 1993, when the Public Broadcasting Service broadcast a five-part documentary titled *Healing and the Mind*. Hosted by the respected television journalist and political commentator Bill Moyers, the series billed itself as a down-to-earth survey of the new claims for mind-body healing.[48] Individual episodes of the documentary explored the effects of group therapy on breast cancer (David Spiegel's therapy was featured); meditation for stress reduction (Jon Kabat-Zinn's program was profiled); holistic healing centers for cancer; and psychoneuroimmunology. But perhaps the most striking feature of the series was its opening episode, "The Mystery of Chi," a view of mind-body healing from China. The story is classic "Eastward journeys" material. Bill Moyers, embodying the common-sense skepticism of the ordinary man on the street, travels to China in the company of David Eisenberg, a respected Western-trained doctor who also previously studied Chinese medicine in China. Eisenberg functions as the cultural broker of the film, the bridge between East and West.

From the opening strains of Asian music and images of temples and palaces, the Orientalist sensibility of the film is made clear. Even though Moyers and Eisenberg are in Beijing, there are no shots of high-rise buildings, government offices, Chinese businessmen in suits, or streets choked with traffic. Instead, we follow Moyers as he sniffs boiling pots of strange-smelling potions, marvels over dried scorpions, talks to wise teachers, asks wondering questions about the elderly Chinese people in the park performing strange exercises (including vigorous men in their nineties), tries to grasp how needles stuck in patient's faces could be doing them any good, and struggles to remain open-minded about the plausibility of a treatment that involves a physician simply waving his hands, Mesmer-like, over the body of a patient.

As the story unfolds, Moyers learns that all of these treatments are grounded in an invisible life force called qi (pronounced "chee"). To be healthy, Chinese doctors tell him, it is essential that one's qi flow freely and strongly. Herbs and acupuncture are one way to strengthen and unblock qi; mental exercises, especially a practice called qigong, are another. Masters of qigong become so good at moving their own qi, Moyers learns, that some of them can direct it out of their own bodies. In the case of martial arts, directing qi outside the body can repel an attacker (the film shows a rather campy demonstration of this). In a medical context, directing qi out of the body and into the body of a patient can act to revitalize his flagging or stagnant qi.

In Moyers's companion book to *Healing and the Mind*, qi is described as a mysterious force, a force that, "as a physical reality . . . makes no sense at all." Moreover, "the fact that the exercise [of qigong] is thousands of years old, and a hallmark of Taoist, Buddhist, and imperial Chinese scholarship, does not necessarily mean that human beings hold within them rivers, streams, and pools of vital energy. Nonetheless," Moyers concedes, "the mystery challenges." In the film Moyers and Eisenberg take various thoughtful walks through the gardens of temples and palaces, and they ponder the best way to rise to the challenge. It is Eisenberg who proposes the classic "Eastward journeys" solution: these Eastern ancient traditions must now submit to the rationality and rigor of Western science. In this sense, East and West should seek to come together and complete each other.[49]

The ideas of qi and qigong were perhaps quite mysterious to the American audiences who watched that first segment of *Healing and the Mind*; but in another sense, they were not unfamiliar. By 1993, Americans of a certain age and disposition had been exposed for some twenty years to the Hollywood-made martial-arts films that played with these concepts in various ways (the first of these, Bruce Lee's *Enter the Dragon*, was released in 1973). In the late 1970s and 1980s, the *Star Wars* films had introduced millions of Americans to "the Force"—described in the first film as "an energy field created by all living things. It surrounds us, penetrates us, and binds the galaxy together." Observant analysts of Orientalist motifs in the *Star Wars* series have noted that one of the original and greatest of the Jedi warriors is called Qui-Gon Jinn.[50]

Were it not for Bruce Lee and *Star Wars*, it is possible that "The Mystery of Chi" would have met a much more uncomprehending or hostile reception. As it was, both the media and the general public proved remarkably receptive, and Bill Moyers was widely credited with awakening the American public to the potential of Chinese mind-body healing exercises. By the mid-1990s, websites had sprung up offering trips to study Chinese medicine "in the style of Bill Moyers." Martial-arts centers added qigong teachers to their staff, books began to be published (a quick search in 2007 on Amazon.com turned up close to two thousand titles), and commercial videos about qigong did brisk sales under titles like "Secrets of China."[51] In 1997, the city and county of San Francisco went so far as to proclaim November 20–26 "Qigong Week." By 2004, following several smaller studies of the practice in regional American hospitals, the National Center for Complementary and Alternative Medicine (a part of the National Institutes of Health) had added qigong to its portfolio of alternative medical practices worthy of scientific investigation.[52] It had little choice: patients were turning to this practice in growing numbers, and testifying to its apparent effectiveness. In 1996, a journalist for the *Los Angeles Times* reported the story of one of these:

> "I have seen a miracle," Adams said recently, slowly measuring his words. Added his wife: "Terry's victory is over the patronizing attitude of Western medicine. Doctors scoffed at qigong, said we could put Terry in a corner and throw chicken blood on him if he thinks it's helping. Now they're all impressed with Terry's progress. But I've stopped caring what they think."[53]

The fact that China emerged in the second half of the 1990s as an intriguing new reference point for Americans in search of unsullied ancient Asian mind-body practices is rife with irony. To understand why, we need to appreciate some of the background history. Consider this: had a journalist like Moyers undertaken his Eastward journey seventy years earlier, he would not have been presented with a picture of China as an ancient mind-body culture. Instead, he would have found a government that, in 1911, had overthrown its dynastic system of rule and was hard on a path to modernization. He would have been told that continuing allegiance to the old traditional healing methods—acupuncture, herbalism, cupping—was actually undermining the health of the people. Above all, he would have learned that it was interfering with government-led efforts to control the epidemic spread of infectious disease by educating the Chinese people in modern germ theory.[54]

What changed everything was the 1949 Communist revolution, spearheaded by Mao Tse-Tung. Initially, Mao had taken the same scornful view of traditional medical practitioners as the republican government, at one point famously comparing them to "circus entertainers, snake oil salesmen, and street hawkers"—a comment that ended up in Mao's Little Red Book, read by millions of Chinese.[55] In the mid-1950s, however, he changed his position. Traditional Chinese medicine, he announced, was to be cultivated and celebrated as a "national treasure." In 1957, new colleges of traditional Chinese medicine were founded in Chengdu, Shanghai, Gaungzhou, and Beijing. The doctors assigned to these hospitals worked with officials to sift through the teachings of a range of previously autonomous schools and sects. Their aim was to extract the most sensible elements in order to create a single tradition stripped of all "feudalistic superstitious" beliefs and made consistent with modern science and Marxist-Leninist principles. "Chinese medicine," Mao proclaimed, "is a grand cache of knowledge that we should actively bring to light and further evolve."[56] This proclamation was soon reproduced on the front page of traditional Chinese medical textbooks everywhere.

It was a classic exercise in the making of a tradition, and there were several reasons why it happened. Mao was looking to reduce China's dependence on the Soviet Union. The Soviet Union was training most of China's doctors and providing most of its medical supplies. By investing in China's indigenous culture of traditional medicine, Mao could distance China from the Soviet Union on both economic and ideological fronts. China's socialism would not just be a carbon copy of the Soviet Marxist-Leninist model; it would incorporate the national traditions and history of China itself. At the same time, because traditional medicine could be practiced largely using local resources, it offered a way to provide basic health care to the desperately underserved rural regions of the country. This thinking produced the famous barefoot doctor system of the 1960s, in which local farmers were given basic training in sanitary practices and a smattering of Chinese and Western therapies. Working in the fields alongside their patients, many of these new doctors grew their own herbs for use in traditional potions.[57]

News of the barefoot doctor movement spread to the West and excited considerable admiration among some left-leaning observers. Nevertheless, there was little awareness at the time of how it worked in detail or of the larger policy and politics behind it. The United States had broken off diplomatic relations with Communist China in 1949, and the Bamboo Curtain had largely excluded visitors from the United States since then. In the early 1970s, however, the Nixon administration began to explore the possibility of restoring diplomatic relations with China. Secretary of State Henry Kissinger took the lead in laying the groundwork for the first meeting between Nixon and Mao. And in July 1971, *New York Times* journalist James "Scotty" Reston, who had accompanied Kissinger on one of his trips to Peking, had an unexpected encounter with the newly fashioned traditional medicine of Mao's China.

It happened like this. While in Peking, Reston suddenly developed acute appendicitis and had to be taken to the local "Anti-Imperialist Hospital" (originally built with money from the Rockefeller Foundation, but nationalized under Mao in the early 1950s). His hosts were good enough to take down a sign that hung over the entrance bearing Mao's warning that the time was not far off "when all the aggressors and their running dogs of the world will be buried," and he was given the most solicitous of care. His appendix was removed using conventional Western techniques, but on the second night after the surgery he was in great pain, and the doctors at the hospital asked him if they could try a different approach to treatment. He granted permission, and they inserted three acupuncture needles into his right elbow and below his knees. They also circled burning herbs over his abdomen. In short order, he experienced a rapid reduction in the distension of his stomach and relief of his abdominal pain.

All this seemed so astonishing that, when Reston returned to the United States, he wrote an article that the *New York Times* published on its front page under the disarming title "Now, Let Me Tell You about My Appendectomy in Peking . . ."[58]

Here, with humor and a certain obvious affection for his Chinese caretakers, Reston told the story of his medical encounter with "the very old and the very new." His doctors, he said, were "quite frank in saying that the sole purpose of their profession since the Cultural Revolution of 1966–1969 is to serve all the people of China, 80 percent of whom live on the land."[59] For many of them, trained originally in Western medicine, the traditional methods had seemed just as strange as they would feel to most Americans, but they had been persuaded to take them seriously because they had tested them, in and out of the clinic, and seen what they could do. Reston continued:

> Prof. Chen Hsien-jiu of the surgery department of the hospital said that he had studied the effects of acupuncture in overcoming postoperative constipation by putting barium in a patient's stomach and observing on a fluoroscope how needle manipulation in the limbs produced movement and relief in the intestines.[60]

The conclusion was clear. This was a piece of the new China to be watched. And people in the United States, then in the grip of the early years of its own alternative and holistic medicine movement, took notice. Reston's story was retold in newspapers across the country.[61] Within months, a small commission of prominent American doctors had accepted an invitation by the China Medical Association to come observe for themselves the ways in which these traditional medical techniques were used. In short order, the commission filed an enthusiastic report in the pages of the *Journal of the American Medical Association*, in which they described seeing not just patients who experienced postoperative pain relief through acupuncture but patients who underwent surgical procedures with nothing other than a few slender acupuncture needles inserted in various parts of their bodies to protect them from pain and shock. One man was described as sitting up immediately after an operation in which a tumor had been removed from his thyroid, downing a full glass of milk, holding up his copy of Mao's Little Red Book and declaring, "Long live Chairman Mao and welcome American doctors." He then put on his pajama shirt and walked out of the operating theater without assistance.[62] In 1972, Nixon's personal physician, Walter Tkach, led a second expedition to determine whether the whole thing was a hoax. His conclusion: "I have seen the past, and it works."[63]

By now, Nixon himself had traveled to China and signed a number of agreements with the Maoist government for cultural exchanges. One of these exchanges brought practitioners of the traditional healing methods to the United States (they went first to San Francisco), and another involved American medical students going to China to learn about its medical system. In 1979 the first beneficiary of this second exchange program, a recent medical graduate named David Eisenberg, became the first American exchange student to study in China since 1949.[64]

It was also in 1979 that the authorities in China made a further move in their construction of traditional Chinese medicine by adding qigong to the arsenal of officially recognized therapies. Qigong (the word itself was coined only in 1953) was a medicalized and secular extraction of a range of diverse practices that had originally been associated more with martial arts, advanced Buddhist meditative practices, and Taoist longevity rituals. Interest in its possible clinical and health-giving properties had been growing in China since the 1950s, when claims began to be made for its ability to cure cancer.[65]

By the early 1990s, China was in the grip of a qigong medical boom. Elderly people practiced it in the hopes that it would extend their lives. China's government-sponsored Qigong Institutes employed clinicians who could allegedly move their own qi out of their bodies and into the bodies of patients (this was called waiqi, or external qi). Surgical procedures were carried out using waiqi alone—no needles required. Studies were carried out to investigate whether qigong unleashed normally dormant paranormal abilities, like telekinesis or clairvoyance.[66] Given all the interest in China about the new medical qigong, it is hardly surprising that, when Bill Moyers asked David Eisenberg to help him discover the mind-body healing culture of ancient China, their attention should have been predominantly directed to these new, highly popular therapeutic practices. Put another way, the enormously influential "Eastward journeys" story about ancient China and qigong that was broadcast on American public television in 1993 was constructed from ingredients that were as much new as they were old, as much products of China's stormy modern history as legacies of its alleged timeless wisdom.

All this matters, because it highlights a fundamental ethical instability at the heart of the "Eastward journeys" narrative genre. In telling "Eastward journeys" stories, we variously look to India, to China, and (as I discuss below) to Tibet to function as our Other. The East, though, is not really our Other, and never was. Therefore, in each story we orchestrate, actors must be found to play the role of ancient wise man or ancient healer. Some of the people we recruit to that role may indeed be wise and may indeed have things to share, but all are also all real people, who come from countries with histories at least as complex as our own. In this sense, ironically enough, "Eastward journeys" stories rarely, if ever, take us into another world; they just take us deeper into ourselves.

On a trip I took to China in 1999, I visited a modern department store in Beijing that was selling a gadget called (in English) "The Magnetic Acupuncture Eye Massager." The device consisted of a set of eye goggles with a strap that was outfitted with a series of soft, inward-pointing spikes designed to stimulate acupressure points on the head. When the device was turned on, the whole apparatus vibrated intensely. The English-language brochure I was given explained why this was done:

> With the development of modern biological magnetic science, plenty of research and study prove that magnetic field can have a great effect on human body's tissue, organs, nervous system, biological enzyme and biological magnetic field. . . . Applying the above theory and combining [it with] the curing theory of channel acupuncture in traditional Chinese medicine, this product uses both magnetic acupuncture and mechanical acupuncture to activate the important acupoints around the eyes and harmonize the blood, *Qi*, thus improving the adjusting functions of the eye muscles and eye

> nerve, and curing nearsightedness and all kinds of eye-related diseases. . . . This equipment is designed accurately by computer to accord to the shape of the eyes and the distribution of acupoints.

I put the gadget on, allowed the demonstrator to flick the switch, and spent a few minutes feeling the acupressure points vibrating all around my head. The experience was a little unpleasant but vaguely impressive. Leaving the store, I had the thought that I had just experienced a perfect minidramatization of an "Eastward journeys" story—but one that was designed to be primarily experienced, not by Americans or other Westerners but by modern Chinese shoppers! Here was a machine that offered the power of qi and explained the effects with a bit of talk about biomagnetic theory (very new and scientific so therefore good) and a bit of talk about channel acupuncture (very ancient and Chinese, so therefore also good). Ancient healing traditions met modern science and technology in that department store in Beijing. I thought: modern Chinese and modern Americans have more in common than they sometimes realize. Both suffer from modernist malaise, both are drawn to "health" and "healing" as an arena for working out cultural and spiritual dislocations, and both are attracted to a vision of ancient wisdom validated by science and updated to appeal to quick-fix consumerist sensibilities.

Buddhist brains: A Tibetan re-enchantment

Meanwhile, public interest in the health benefits of sitting meditation remained high, even as the cultural resonance around it for many people began, once again, to shift. The original medicalization of meditation in the late 1960s and 1970s by people like Benson and Kabat-Zinn was a self-consciously secular effort that aimed to distance the practice from all its exotic Eastern associations. By the 1990s, however, there began to be signs of a change afoot on this front, of a turn back to the East. The face of the East to which people turned this time was not India, not China; it was Tibet, perhaps the most exotic of the three.

The origins of these developments lie in 1979, when the Dalai Lama of Tibet, the political and spiritual leader of the Tibetan people (in exile since the late 1950s), made his first visit to the United States: a forty-nine-day tour to cities, houses of worship, and university campuses across the country. It played as a thoroughly romantic and exotic event, and media coverage was intense.[67] It was intense in part because Tibetan Buddhism itself was still a mysterious and largely unknown tradition to most Americans. In fact, when some of the Dalai Lama's American followers called up television programs in 1979 to ask if they would care to interview him, a common response was, "What did you say her last name was?"[68]

In 1979, Tibet's political plight was also far from the popular cause it later became. Although officially opposed to the occupation of Tibet, successive U.S. presidents had grown increasingly reluctant to challenge China too explicitly, not wanting to destabilize the fragile foundation of goodwill laid down by the Nixon administration. For this reason, even though the Dalai Lama had made it known for some years that he wished to visit the United States, the U.S. government had found various excuses to deny him a visa.[69] When in 1979, under the Carter administration, the visa was finally granted, it was on the strict condition that the trip be completely apolitical. The Dalai Lama was coming to the United States not as an exiled head of state but simply as a religious leader on a spiritual visit. There must be no discussion of the Chinese occupation, he was told, or anything else politically sensitive.[70]

The Dalai Lama was as good as his word. Instead of talking politics, he engaged in discussions about Buddhism, Western culture, different religious traditions, and ethics. He also signaled an interest in talking about Western science—a personal fascination of his that went back to his boyhood (he was fond of saying that if he had not become the Dalai Lama, he would have liked to be an engineer). "I'd like to listen to your experiences," he told a panel of scholars in Texas who had gathered to engage with him. "I'd like to hear what is the latest in research into the relations between consciousness and matter, and how they affect each other."[71]

In fact, the scholars in Texas did not have much to say about those matters, and so the conversation there turned in other directions. Two weeks later, however, the Dalai Lama came to Harvard University for a three-day visit, and one of the people he met there was Herbert Benson, who did have things to say on this topic.[72] Benson told the Dalai Lama all about his research on stress and the relaxation response and how he believed it cast light on the physiology of meditation. Then he admitted that he had a request of his own. He knew that there existed an esoteric Tibetan meditative practice called g Tum-mo (pronounced "tummo"), or "inner heat" meditation, during which advanced practitioners were said to be able to regulate their heat production in ways that allowed them to stay warm even in frigid weather conditions and without warm clothing. If true, this would be of great physiological, and possibly clinical, interest. Would the Dalai Lama assist him, Benson asked, in the task of persuading appropriate Tibetan monks to participate in some physiological experiments?

As Benson later publicly recalled the conversation, the Dalai Lama's response was immediate and straightforward: "His Holiness agreed to help me."[73] Jeffery Paine, in his book on the rise of Tibetan Buddhism in the West, tells a different story about that same encounter. He points out that normally it would be unthinkable for a Buddhist religious leader to agree to any kind of unnecessary interference with the spiritual practices of his community, and he noted that indeed the Dalai Lama first began to explain in Tibetan (through a translator) why he must refuse Benson's request.

Then, according to Paine, he stopped, thought again, and suddenly broke into English: "Still," he said, "our friends to the East [meaning the Chinese] might be impressed with a Western explanation of what we are doing." And in that moment, he decided to break "a millennium's worth of precedent" (in Paine's words) and told Benson that he would help him study the monks. The Dalai Lama had not been allowed to be overtly political during this trip to the United States, but he was not going to turn down an opportunity to improve the image of Tibet and its religious culture in the West and China. As the spiritual leader explained to a group of Tibetan monks soon after, "For skeptics, you must show something spectacular because, without that, they won't believe."[74]

Benson's investigations of g Tum-mo meditation did indeed resonate with long-standing Western interest in what we might call the "spectacular" face of Asian spiritual practices. Could yogis really stop their hearts, as one French researcher from the 1930s had claimed? Could others survive for extended periods locked up in airtight boxes, buried alive, or as in this case, while meditating all night in the Himalayas, under ice-cold conditions that would kill a normal person?[75]

It was Benson's background in biofeedback research that made claims like these feel more plausible to him—and others—than they might otherwise have been. If quite ordinary people, assisted by feedback from electrophysiological instruments, could learn to control normally involuntary bodily processes, then it was at least possible that yogis had simply mastered these same processes of self-regulation, but to a degree that had no real analogue in the West. That is to say, perhaps it was possible to think of yogis as super biofeedback self-regulators. Not only could they take the processes of autonomic regulation much further than normal Western patients, but they didn't need machines. Even before Benson approached the Dalai Lama, another first-generation biofeedback researcher, Elmer Green, had been intrigued enough by similarities between the results of biofeedback and the age-old claims about the abilities of yogis that he had run tests of a high-profile Indian yogi teacher named Swami Rama who claimed, among other things, to be able to stop his heart from pumping blood. In fact, Green said, he could: not by actually stopping his heart, but by speeding it up to such a high rate that he produced temporary atrial flutter.[76]

Now Benson had his own opportunity to see what substance, if any, there might be in the spectacular claims of yogis. Initially, he made several trips to the area around Dharamsala, India, to make informal observations and measurements. There, he first was able to document significant increases in the temperatures of the fingers and toes of several individual monks versed in the practice (the ambient air temperature also increased modestly during the period the monks were meditating, leading some critics to call the results into question).[77] In 1985 Benson's team returned to India and were able to take video footage of monks who, in ambient temperatures of 39 degrees Fahrenheit, allowed sheets dipped in cold water to be draped over their shoulders and then tried to dry them; the film seems to show steam rising up from the sheets.[78] Another, more formal study based in Normandy, France, was inconclusive, as was an effort in which three monks traveled from India to Benson's laboratory in Boston.[79]

Be that as it may, the Dalai Lama, with his precedent-changing decision to allow Western scientific research into the physiology associated with Tibetan religious practices, had opened a door to a new and distinctly exotic phase of research into meditation. Though it would take

Benson's early investigation of Tibetan monks (living in Dharamsala, India) who practiced the advanced contemplative practice in Tibetan Buddhism called g Tum-mo (pronounced "tummo"). The photo indicates that the date of this test was February 22, 1981. Among other things, g Tum-mo is supposed to generate intense sensations of body heat, even under frigid conditions. *Courtesy of the Harvard Medical Library in the Francis A. Countway Library of Medicine*

another decade, by 1990 others besides Benson would begin to knock on that same door.

It happened like this. In 1990 a young organization called the Mind and Life Institute arranged for the Dalai Lama to spend five days in his home in Dharamsala talking with clinicians, researchers, and religious scholars on the topic of "healing emotions." Mind and Life had been cofounded in the mid-1980s by the Chilean neuroscientist Francisco Varela (well known as the codeveloper with Humberto Maturana of the concept of autopoiesis, the idea that living systems are self-organizing entities)[80] and the American businessman Adam Engle. Mind and Life was committed to facilitating dialogue between the Dalai Lama and Western scientists (today, it has expanded its mission to also include research and outreach),[81] and the first meetings, spearheaded by the philosophically minded Varela, had focused on developing the principles for a coherent and sensitive dialogue between neuroscientific and Buddhist approaches to investigating the mind.[82] The more practical, explicitly clinical focus of the "healing emotions" conference represented a new direction for Mind and Life.

In this meeting, questions were asked about what was known about the physiology of meditative states, and the answer, predictably enough, was: not enough. As the conversation continued, both in the meeting and informally over meals, a new idea emerged. Perhaps, with the advent of new, more portable instruments for measuring cognitive and brain function, it was now possible to do the kinds of studies of which people previously had only dreamed: to study the effects of long-term and intensive meditative practice on the brains and cognitive functioning of senior monks from the Tibetan monastic community.[83]

Not only was the Dalai Lama enthusiastic about the idea when it was presented to him, he actually went to the trouble of personally asking the Council for Religious and Cultural Affairs of the Tibetan government-in-exile to identify the most experienced senior monks living in retreat in the mountains above Dharamsala, so he could provide that information to the scientists who would do the research.[84]

Thus it came about that, in 1992, a small expedition of scientists accompanied by two Western-born Tibetan scholars (both fluent in Tibetan) traveled to Dharamsala to discuss a proposal for a research effort with ten senior monks. To their surprise, and in spite of the strong support of the Dalai Lama for this work, the Westerners met considerable resistance. Some of the monks were dubious that the research had any clear ethical value and therefore thought it unlikely to be worth their time. Others worried about possible personal risks they might be taking by participating. Still others wondered how a study of their meditative practice that was focused on physical measurement could make sense; they had no doubt that the mind was nonphysical in nature. In addition, since the study did not take into account the doctrine of reincarnation—a fundamental rationale for Buddhist meditative practice—how could the scientists hope to make proper sense of any results they might obtain?[85]

In the end, this effort was aborted, but it left a legacy: it helped reawaken partially dormant, but deep-seated, ambitions of one of the scientists involved in that original expedition to bring meditation squarely into the mainstream of psychology and brain science. In 1990, Richard Davidson was a professor of psychology at the University of Wisconsin in Madison. He had spent more than a decade studying asymmetries in hemisphere functioning and their relationship to affective experience; but at the same time, he was nurturing an interest in meditation that went back as far as the early 1970s when he was a graduate student at Harvard.

Current photograph of Richard Davidson, Ph.D., University of Wisconsin, Madison. *Courtesy of Jeff Miller / University of Wisconsin, Madison*

There, at the height of the first counterculture romance with all things Eastern, he had become friendly with Daniel Goleman and Jon Kabat-Zinn (both also deeply immersed in the study and practice of Asian spiritual traditions at that time); he had spent time with such Harvard counterculture icons as Ram Dass (formerly a social psychology professor named Richard Alpert); and he had even taken an extended break from his graduate studies in psychology in order to travel to India for more advanced personal training in meditation.

In their time as students and young researchers together, Davidson and Goleman had worked through publications, collaborative research undertakings, and other activities to put meditation on the map of experimental psychology.[86] Nothing much had come of those first efforts. But meeting the Dalai Lama, encountering the Tibetan monks on the hillside above Dharamsala, and becoming involved with the Mind and Life Institute: all of these things worked to convince Davidson that perhaps the time was right to try again.

Helpfully, Americans in these years were in the middle of a growing love affair with Tibet in general and the Dalai Lama in particular. In 1989, the Dalai Lama was awarded the Nobel Peace Prize for his nonviolent efforts to negotiate the Tibetan cause with the Chinese government. In the years following, he traveled ever more frequently to the West, where his irrepressible good humor and charismatic spiritual presence helped give new vitality to older romantic visions of Tibet as a Shangri-la (now politically threatened), with much to teach a morally and spiritually impoverished West.[87] Everyone, it seemed, wanted the opportunity to explore the deep secret behind this man's happiness and equanimity, and to bask in his warmth.

Moreover, word was now out that this man who embodied ancient wisdom was also interested in science—especially, it seemed, the sciences of mind and brain. Scientists attending the private Mind and Life meetings in India in the 1990s were proud to show the spiritual leader their vivid images of the brain in action and to tutor him in the latest developments in their fast-changing field. Early in the decade, he was feted at a conference at Harvard Medical School, cosponsored by Tibet House, where Herbert Benson presented him with the first results of his research on g Tum-mo.[88] In May 1998, he was the guest of honor at a public conference in New York City that brought together a star-studded cast of brain scientists, physicians, and Buddhist scholars to discuss the "health effects of advanced meditation."

This last event is worthy of further comment, since it provided a foretaste of things to come. *New York* magazine's cover that week sported an image of the Dalai Lama posed in front of rows of computers showing colorful brain scan images. Dressed in traditional monastic robes, with hands clasped and a serene (if slightly puckish) expression on his face, the Tibetan leader had an array of EEG electrodes pasted across his head. "Spurred by new developments in neuroscience and by hefty donations from believers—not to mention a visit from the Dalai Lama," *New York* magazine announced, "New York's straitlaced medical Establishment is finally recognizing the healing power of mind-body techniques like meditation."[89]

Several months later, the Buddhist magazine *Tricycle* published a brief article about the conference that buzzed with talk about bio-individuality, psychoneuroimmunology, quantum consciousness, and relativity theory. The encounter between brain science and the Dalai Lama in New York was, the readers of *Tricycle* learned, part of a larger quantum " 'wave revolution' that [was] shifting Western science toward the wisdom of Asian traditions." Indeed, the article concluded rather breathlessly, it was more than that: in New York City that day, "Arnold Toynbee's vision of a world renaissance sparked by the meeting of East and West seems to have arrived."[90]

I was an observer at that New York City meeting. What struck me at the time was less a sense of radical East-West convergence than a distinct lack of thematic integration. The conference participants discussed, on

Ancient Eastern wisdom meets modern Western neuroscience? The Dalai Lama gazes out from the cover of *New York* magazine with his bald pate covered in electrodes and posed against a bank of computers. The scene is actually a cultural fantasy image: the *New York* magazine photojournalist has reworked a standard photograph of the Dalai Lama in a way that has the effect of literally draping him in symbols of brain science research. *Permission courtesy of* New York *magazine. Photo illustration by John Blackford, 1998. Photographs: Eddie Adams@Corbis/Outline (Dalai Lama) and FPG (computer)*

the one side, their laudable but arguably low-tech goal of finding ways to teach meditation to ordinary patients in hospital settings to reduce their suffering. The Dalai Lama was then regaled with a stream of presentations by scientists about various high-tech methods in the brain sciences. At the same time, Buddhist scholars talked about Tibetan understandings of consciousness and whetted other participants' appetites to learn more about the highest and most esoteric states of meditative consciousness. For some, it was like the 1960s all over again.

Indeed, the fascination with the exotic was so palpable in this meeting that at one point the Dalai Lama pointedly noted that, of the six levels of meditation being discussed, he himself had experienced only the first three. He then gently suggested that the participants might get further in their practical clinical efforts if, rather than trying to figure out what "bliss" was, they focused instead on helping hospital caregivers cultivate greater reserves of kindness toward one another and their patients.

Nevertheless, since 1998 it has been "advanced" meditation—the meditation accomplishments of exceptional and often exotic-seeming people—that has tended to be at the forefront of conversations about the general benefits of meditation. More than anyone else, Richard Davidson has taken the lead in setting this emphasis. It is true that, in the late 1990s, Davidson's laboratory had run tests on a group of ordinary employees who underwent Kabat-Zinn's eight-week MBSR training, finding some evidence that participating in this program (though not necessarily meditating regularly) shifted the emotional set-point in the brain and enhanced the responsiveness of people's immune systems.[91] Davidson had never abandoned his interest, however, in putting the brains and minds of the most advanced monastic practitioners through their paces in the laboratory to see what he might find. His efforts in the early 1990s had fallen short, but by the end of the decade the situation was quite different.

For one thing, in part through his connections with the Mind and Life Institute (on whose board he now sat), Davidson now had a partner, a French-born Tibetan Buddhist monk named Matthieu Ricard, who for many years had served as a translator for the Dalai Lama. Ricard was no ordinary Tibetan monk. Before he had entered monastic life thirty years earlier, he had been a graduate student in biochemistry in Paris, working under the Nobel laureate François Jacob. In the intervening years, Ricard had published books that established him as a sophisticated interlocutor of Western philosophical and scientific traditions (including a best-selling dialogue with his own father, philosopher Jean-François Revel, published in English under the title *The Monk and the Philosopher*). Now, in the late 1990s, he had learned of Davidson's interest in returning to the study of advanced meditative practice, and agreed to help by offering himself and indeed his very own brain to the cause.[92] He had no fear of science, and he believed that laboratory studies of meditation could only serve to enhance people's interest in this vital practice.

The initial results were better than anyone might have dared hope. When the researchers in Davidson's laboratory put 128 electrodes on Ricard's head and asked him to meditate on "unconditional loving-kindness and compassion," they noticed powerful gamma activity—brain waves oscillating at roughly 40 cycles per second in a highly synchronized

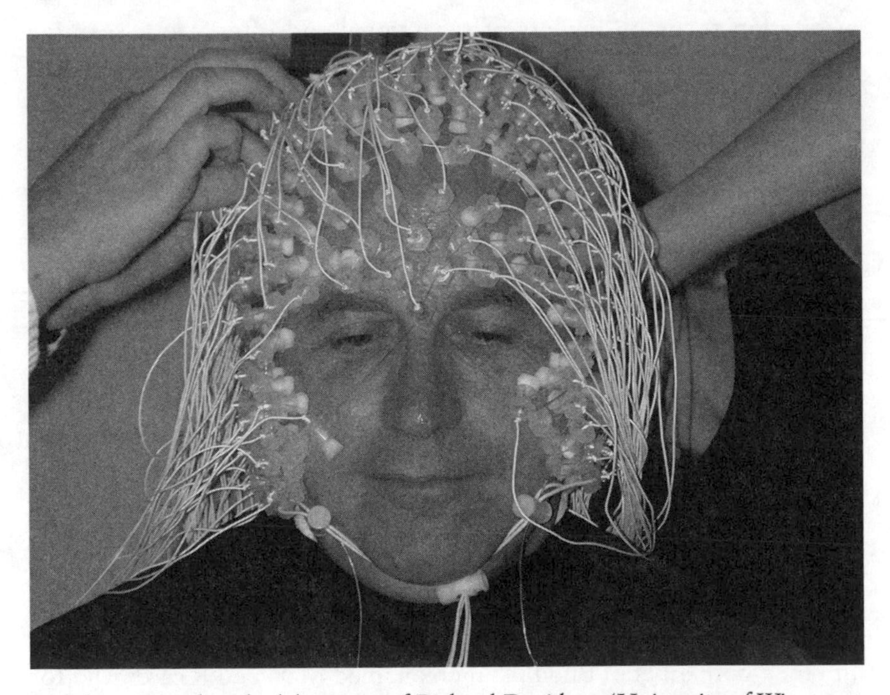

Matthieu Ricard in the laboratory of Richard Davidson (University of Wisconsin, Madison) preparing to undergo EEG investigations of his brain while meditating. *Courtesy of the Waisman Brain Imaging Laboratory, University of Wisconsin, Madison*

fashion across the cortex (this was a phenomenon sometimes seen in patients under anesthesia, but not in waking states). The researchers had never seen anything like it. To be sure there was no mistake, they brought in other long-term Buddhist practitioners (both monks and laypeople) and compared their brain wave activity in meditation to that of a control group of college students inexperienced in meditation. Those practitioners produced gamma waves that were thirty times as strong as the students'. In 2004, Ricard was a coauthor on a paper reporting these and related fabulous-seeming findings that was published in the *Proceedings of the National Academy of Sciences*.[93]

Even before this work was published, Davidson had presented preliminary data that had created considerable buzz. Indeed, this work inspired a 2003 *New York Times* article that asked the telling question "Is Buddhism Good for your Health?" What did monks have to teach the rest of us, the article asked, about optimal functioning?

> The "Monk experiments" at Madison are beginning to intersect with a handful of small but suggestive studies showing that Buddhist-style meditation may have not only emotional effects but also distinct physiological effects. That is, the power of meditation might be harnessed by non-Buddhists in a way that along with reducing stress and defusing negative emotion, improves things like immune function as well.[94]

Two years later, in 2005, Davidson took the lead in persuading the Society for Neuroscience to issue an invitation to the Dalai Lama to speak before its body about meditation and its many benefits. There was a flurry of consternation and, among some, outrage when the invitation was made public, and a petition was circulated demanding that it be withdrawn. Significantly, perhaps, most of the signatories were Chinese or of Chinese origin, but they insisted that their concerns about the lecture were purely scientific. In the words of Yi Rao, a Chinese-American neuroscientist at Northwestern University in Chicago who helped to draft the petition, meditation is "a subject with hyperbolic claims, limited research and compromised scientific rigor."[95] Other scientists simply insisted flatly that religious perspectives should not be given a platform at scientific meetings, regardless of the merit of the topic. "Who's next," snapped one professor of anesthesia, "the pope?"[96] Nevertheless, when it came time for the Dalai Lama to speak, more than thirteen thousand neuroscientists crowded into a huge auditorium in Washington, D.C., to hear him. There he talked about Buddhist meditation and its benefits, and then affirmed his belief in the neuroscientific basis of those benefits. The scientists gave him a standing ovation.

As I write, the history of this particular "Eastward journeys" narrative continues and it remains a place to watch. There is perhaps no other place in contemporary American mind-body medicine where extraordinary personalities and high-profile moments dominate like they do here; no other place where questions about cultural significance—rather than just scientific and medical validity—matter more.

Conclusion

Making Sense of Mind-Body Medicine

We live in a world where scientists have mapped the human genome, where the pharmaceutical industry is a multibillion-dollar business, and where people look to the brain sciences to illuminate everything from schizophrenia to shyness. Reductionist medicine, it would seem, has triumphed. Nevertheless, this book has been all about demonstrating that there is more to say much more. In March 2007, I conducted an internet search on Amazon.com using several keywords. Amazon showed me 2,438 books that were about the "mind-body connection" and 1,081 books that were about "mind-body medicine." It also showed me 1,992 books about psychoneuroimmunology; 1,638 books about social support and health; 1,717 books about the health benefits of meditation; 5,913 books about the placebo effect; 11,724 titles concerned with stress and health; and a whopping 14,075 titles concerned with positive thinking. When I searched on the keyword "genomics," 15,138 titles turned up. Genomics is obviously important in our culture—we knew that—but mind-body medicine appears to be holding its own.

What accounts for this? There are at least four related factors that, taken together, begin to help us make sense of the kinds of numbers we see here. The first of these, ironically enough, has to do with the fact that mind-body medicine—a term that came into common parlance only in the 1990s—is not a single approach to healing, but a recently constructed patchwork of quite distinct narrative traditions: the narrative traditions whose histories I have traced in this book. In the past, these traditions have sometimes interacted cooperatively, and sometimes more antagonistically. They were never, however, seen as all part of the same grand enterprise. Stress was never the same thing as psychoanalysis. Meditation research had little, if anything, to do with social epidemiology.

Even when efforts were made in the past to assert the need for a more broadly based programmatic vision, it was different than what we see with mind-body medicine today. "Psychosomatic medicine," the synthetic program favored in the 1940s and 1950s, signaled a very specific programmatic commitment to a Freudian or at least broadly psychoanalytic approach to mind-body therapeutics—to ideas about speaking bodies. "Behavioral medicine," a movement that arose in the 1970s, stood above all for a commitment to interdisciplinary science, with a particular emphasis on laboratory-based rigor. "Holistic medicine," also coined in the 1970s, functioned above all as a political statement of commitment to healing the whole person in clear opposition to an increasingly vilified reductionistic medicine. The new mind-body medicine sent a different message: one no longer needed to choose a camp.

More than anything else, it was Bill Moyers's 1993 five-part series *Healing and the Mind* that served to focus public attention on this new mix-and-match vision of mind-body healing. On consecutive nights, viewers were introduced to discussions of ancient Eastern healing practices, meditation, group therapy, the healing power of community, stress reduction, placebo healing, and various perspectives on the science behind it all. The initial broadcast by the Public Broadcasting Service was viewed by more than twenty-four million Americans, and its Doubleday companion book stayed on the bestseller list for twenty-three weeks.[1] After the show aired, PBS's phone lines were flooded with people wanting more information about the practices shown; bookstores began receiving queries from people looking for books on mind-body therapies; the offices of unconventional medical practices began to receive calls from prospective patients who would not have previously considered these alternatives to orthodox medicine.[2] Journalists later gave the series credit for the consolidation of mind-body medicine as an independent self-help publishing category, as well as for the upsurge of interest in previously obscure healing practices such as qigong. As one reviewer summed up the emerging public consensus, "The mind-body connection must be OK—it's on the Bill Moyers show."[3]

The year 1993 also saw the publication by Consumer Reports of a guide coedited by Daniel Goleman and Joel Gurin: *Mind Body Medicine: How to Use Your Mind for Better Health*.[4] Originally envisioned as a companion volume to *Healing and the Mind* (and financed by the same organization, the Fetzer Institute, that partly funded the Moyers series), it helped to solidify "mind-body medicine" as the new preferred term of choice for this approach to healing.[5] By the mid-1990s, the term was being used in the titles of both paperback best-sellers advocating radical New Age visions of mind-body healing and political hearings in a subcommittee of the United States Senate.[6]

Consider the consequences. Today's mind-body medicine offers resources for proponents of doctor-led rituals who may also be skeptical of patients' own abilities to control and make sense of their own experiences ("The Power of Suggestion"); for those who believe in the healing power of the examined life ("The Body That Speaks"); for advocates of patient-initiated practices and those most skeptical of medicine's arrogance ("The Power of Positive Thinking"); for those most committed to the power of modern laboratory science to crack the secrets of the mind-body connection ("Broken by Modern Life"); and for those who are drawn to both the more folksy and homegrown ("Healing Ties") and the more exotic and romantic ("Eastward Journeys") forms of medical, social, and moral redemption.

It is not just on a practical, therapeutic level, however, that mind-body medicine appears to offer something for everyone. This book has been all about the extent to which mind-body medicine also needs to be appreciated as a set of narratives that offer people a diverse set of cultural resources to make sense of their experiences. Many of these, as we have seen, have strong historical origins in religion. The links to religious understandings and practices are the second factor that helps explain the strong hold that mind-body medicine has on our cultural imagination. Thus, today's visions of minds in the thrall of suggestion emerged out of older visions of bodies possessed by demons, and this distinctive religious origin gives those visions a different, and more dangerous, kind of resonance than they might otherwise have. Similarly, we understand better the persuasiveness of our contemporary beliefs in the therapeutic power of speaking dangerous truths out loud—during hypnosis, or lying down on a couch—if we recognize that they evolved in part out of ancient religious beliefs about the healing power of confession, including the forced confessions of a demon being exorcised by a priest. There may be better reasons than we sometimes realize for why people sometimes talk about "exorcising their demons" in psychotherapy.

More generally, what we learn from tracking the translation of religious ideas about mind-body healing into the secular idioms of mind-body medicine is this: a belief or practice that is a secularized version of an older religious tradition is not the same as a belief or practice that never had any kind of prior moral or religious meaning. This is because we can almost always discern an echo of the original religious message in the new secular story. Indeed, sometimes that echo (perhaps because it is not officially recognized or acknowledged) can sound more loudly and insistently than the original from which it has derived. A student of mine, who had spent part of his childhood in India, told me that he had been struck by the fact that millions of people in India practice yoga on a regular basis, and do so in a very matter-of-fact way. In the United States, in contrast, where yoga is secularized and widely seen as a health practice, people seem to feel that it can't be effective unless they burn incense and light candles.

The religious roots of some of mind-body medicine's most striking narrative themes account for some of its extraordinary success in our time. But another—the third of the factors with which I believe we must reckon—is this: the fact that, during the post–World War II period in particular, large parts of mind-body medicine have functioned as amplifiers of a range of very distinctive moral and social concerns about the costs of modernity. This has been true particularly in the United States, where debates about a number of emergent chronic and unexpectedly intractable illnesses—stress-linked disorders, heart disease, cancer, AIDS—have interfaced repeatedly with powerful cultural anxieties: about the costs of postwar material prosperity; about the meaning of manliness in the brave new world of the office; about the breakdown of the family; about the dissolution of community; and about discontent with materialistic American values more generally.

The medicalizing of the moral and political in this context not only enhances the perceived relevance of various narrative traditions from mind-body medicine to broader debates; it also gives those debates themselves a weightiness they otherwise might not have. We live in a world today where the wages of (modernist) sin have been understood literally (at least, for some) to be disease or death; a world where lonely and uprooted Americans are not just figuratively but literally brokenhearted; a world where the decision to cultivate the simple virtues of home and hearth—or, alternatively, to humble one's self before the superior wisdom of some Other—is believed to translate into longer life, better immune functioning, and more flexibly functioning brains.

The eclecticism of its current programmatic vision, the longer-standing depth of its religious connexions and connotations, and the extent to which it has historically resonated with postwar America's most prevalent moral and social anxieties all help explain the rise, spread, and persistence of mind-body medicine. But there is a fourth and final factor that must also be taken into account: namely, mind-body medicine's extraordinary ability to flourish simultaneously at many different levels in our culture. One of the most interesting features of mind-body medicine is its unstable status as both a mainstream/professional and an alternative/popular body of knowledge and practice. In no instance have any of its traditions emerged as self-contained bodies of expert knowledge that then perhaps have found some kind of separate and subordinate popular representation. Rather, all of the stories of mind-body medicine have been fluid and elastic things, constantly on the move between one social context and another, and always available for appropriation and reappropriation by different social groups. At particular moments, principles and practices have moved from elite groups of specialist practitioners to mass audiences, and back again in the opposite direction.[7]

This fact helps us make sense of the cultural staying power of mind-body medicine in two ways. In those times when science appears to bless popular interest in a particular mind-body narrative, there can be a process of mutual buttressing and amplification: simply put, scientists can gain credit for being relevant; and at the same time, popularizers can gain advantage for being seen as rigorous and science based. Conversely, in those times when a particular narrative tradition is neglected scientifically, or explicitly falls out of favor, continuing popular interest may serve to preserve for it a place in the culture more broadly, from which it can then be reclaimed by a later generation of scientists.

Thus, mesmerism began as an Enlightenment-era ritual of professional healing based on Newtonian principles. In the early-nineteenth century mesmerism fell out of favor in professional medical circles, but this did little to prevent its growing presence in the wider culture as a form of popular stage entertainment; and at a later stage, it was thus available to be absorbed back into mainstream medical culture under the guise of hypnosis. Similarly, ideas about how to realize the healing effects of positive thinking were first developed and disseminated in populist teachings and texts; next, they were claimed by reform-minded medical researchers in the second half of the twentieth century (who linked these ideas to emerging understandings of the biochemistry of placebo effects and the ways in which the brain and immune system interact); and finally they were disseminated again into the popular culture, but stamped now with the imprimatur of biomedical science. Again, a popular culture of meditation in the 1960s that saw it as a way to experience exotic forms of consciousness inspired a medical culture of meditation in the 1970s and 1980s that emphasized its universal physiological and clinical benefits; and that medical culture in turn merged with a new wave of popular fascination with Tibet in the 1990s to create a new hybrid vision of meditation as an exotic technology for both moral and medical improvement.

In recent years, one of the most important mediators between elite and popular, mainstream and alternative versions of different narrative traditions of mind-body medicine has been the therapeutic self-help industry. To understand why, it is worth having some sense of the longer and complex historical relationship between mental healing practices and self-help. There is an established historiography that sees self-help as one expression of a secular and psychologically minded "therapeutic culture" in general that first arose at the turn of the twentieth century as part of a new American culture of narcissism, consumerism, and consumption—a culture that valorized "feeling good."[8] At the same time, a new generation of scholars also increasingly appreciates the extent to which many practices of modern self-help, particularly as a print culture, owe a significant debt to the religious mind-cure movements of the late nineteenth century.[9] A number of early-twentieth-century New Thought books sold over one million copies. Some of the magazines published by these same groups had weekly subscriptions exceeding 100,000. By turning self-help into a portable print product that could be inexpensively produced and sold on a massive scale, the original mind-cure movements pioneered an outreach strategy that is still favored by the industry today (even as television and the internet have further extended its reach).[10]

There is, though, an important difference between the self-help bestsellers of the mind-cure teachers and most self-help books today. The mind-cure tradition of self-help was essentially a tradition of writing by laypeople for laypeople: the self-made man or miraculously healed woman put pen to paper to tell the world how he or she did it. Since the end of World War II, however, this populist strand of self-help has been increasingly superseded by a new kind of therapeutic self-help culture in which scientific experts trained in psychology or medicine tell everyone else how to live happier, healthier, and more productive lives.[11] Thus Dr. Spock's blockbuster book *Baby and Child Care*, first published in 1946, translated Freudian principles into do's and don't's for millions of anxious parents. In 1953, *Ladies' Home Journal* launched its magazine column "Can This Marriage Be Saved?" and introduced a formula of self-help writing in which readers first hear from a couple with marital difficulties (but little insight as to why), and then hear how professional therapists make sense of the couple's problems and what they believe can be done to save the marriage.

By the 1950s, a newly self-conscious scientific culture of mind-body medicine would adapt increasingly to the norms of this new postwar culture of expert self-help. Norman Vincent Peale arguably marked the launch of this second chapter in the self-help tradition of mental healing with his best-selling 1952 *Power of Positive Thinking*. Peale was, of course, a preacher and not a medical doctor, but he was a mainstream authority—a man who spoke to thousands from a bully pulpit—and he also took great pains to ground his arguments repeatedly in what he insisted was the latest in psychological and medical research. The subsequent emergence in the 1960s of stress as a new, laboratory-based understanding of the sources of human ailing offered a further opportunity for scientific experts (or writers who relied on the authority of those experts) to find a niche for themselves in the self-help publishing world. A 1981 Institute of Medicine report on stress was right to note the "thriving industry" of self-help programs and best-selling books targeted to stress-linked disorders, especially hypertension.[12] The authors of this report almost certainly had in mind best-sellers by medical authorities such as Meyer Friedman and Ray Rosenman's 1974 *Type A Behavior and Your Heart*, and Herbert Benson's 1975 *The Relaxation Response*.

With so much going on, so many mind-body narratives serving so many functions, there comes a moment when we want to say: But which of them are true? My students generally reach this point sooner than I probably would prefer. History is all well and good, they say; but they live in the twenty-first century, surrounded by these ideas, and they need to be able to make some judgments. Are versions of stories that flourish on the kooky alternative margins of society to be treated with the same respect as versions of stories that are grounded in impeccable laboratory research? Are all existing narrative traditions equally worthy of their attention? Is the old-fashioned psychoanalytic tradition—their grandparents' version of mind-body healing—really as worthy of their attention as the new work on neuroplasticity and meditation? A little guidance, please, professor!

The students have reason, perhaps, to believe I may be holding back. Many of them are aware that I have not only been interested in the history of mind-body medicine; I have also spent more than a decade as an engaged witness and sometime contributor to its debates. For example, I not only accept the reality of the placebo effect, but I have argued in several publications that it is a biocultural phenomenon with important lessons for mainstream medical practice and theory. In the 1990s, I spent six years as a member of a research group studying mind-body interactions that was funded by the MacArthur Foundation. Since the second half of the 1990s, I have also been involved in efforts to facilitate dialogue between Buddhism and the neurosciences, through my involvement with the Mind and Life Institute; and I have taken research trips to China to see what I could learn about the potential relevance of traditional Chinese medicine and qigong for Western understandings of mind-body interactions.[13]

I have done all these things because for years I have believed in the potential of mind-body medicine to destabilize and creatively redraw tired "two-cultures" approaches to knowledge of what it means to be human. For years I have been intrigued by the tendency of many universities to put the humanities buildings and the science laboratories on opposite sides of the campus. I have wondered whether the habit of drawing campus maps in this way is really just a matter of rational real-estate planning. Could it be, instead, a reflection of the way in which many academies have wanted to think about the relationship between the natural and the human worlds? In other words, do our literal campus maps function also as metaphorical cognitive maps, and do these cognitive maps serve us as well as they should? Perhaps they do when it is a matter of studying igneous rocks as opposed to Shakespeare's sonnets. But what about when we come to study human beings themselves? Do we humans really divide neatly in two, as so many academic campuses seem to imply: "This way for the body, please, that way for the mind"?

It was in the course of asking questions like these that I first became interested in what I came to think of as bodies that behaved badly; that is, bodies that insisted on having experiences that looked a bit like biology and a bit like biography, a little like mechanism and a little like meaning. In the introduction to this book, I told the story of sitting in a library in Paris and realizing that the history of hypnosis did not just involve changes in ideas and theories over time; it also involved changes in the sorts of experiences people had. This presented a big challenge to me as an historian trying to work out what it would mean to insert the body inside history; but it was also clear that it presented no less a challenge to medical scientists, who rarely if ever think they might need to make room for culture and history inside the body.

Over the years, I began collecting other examples of bodies behaving badly. In no particular order, some of the cases that have haunted me most over the years are:

1. Children living in institutional settings whose material needs are all met, but who nevertheless are physically stunted and developmentally retarded because they lack a secure bond with a loving caretaker;[14]
2. Mortality statistics showing that death rates in particular social groups dip below expected levels just before a symbolically meaningful occasion, such as a major cultural or religious holiday (the Harvest Moon festival for the Chinese, Passover for Jewish males);[15]
3. Some two hundred cases of blindness in a group of Cambodian women forced by the Khmer Rouge to witness the torture and slaughter of those close to them, particularly their menfolk. Examination of these women confirmed that there was nothing physically wrong with their eyes. Their own understanding of what had happened to them was that, having been made to bear witness to the unbearable, they had all "cried until they could not see";[16]
4. The case of Mr. Wright, with which I opened this book: the man whose tumors "melted like snowballs on a hot stove" in response to the "worthless" cancer treatment Krebiozen.[17]

I came into mind-body medicine hoping I had found here an enterprise that had committed itself to studying these badly behaving bodies—and, through that effort, to finding alternatives to our fractured approaches to our humanness. I became involved in a range of projects and came to know some of the players. Finally, and most importantly, I learned the stock narratives of this community—heard them told and enacted over and over by my colleagues. I was struck by what a "storied" world I had entered. I wondered where the stories had all come from, and what roles they were serving. That process of wondering took me back to my home discipline, history, to try to find out.

At the same time, I also had my own opinions about the science I saw on offer. As already mentioned, I studied the literature on the placebo effect and found broad and deep evidence for taking it seriously as a psychobiological phenomenon, especially in the area of pain.[18] I was also impressed, and somehow morally touched, by the range of early evidence that cultivating social ties could affect one's susceptibility to illness and even death. I have recently watched with a certain sense of wistfulness as at least some subsequent efforts to test the therapeutic efficacy of such ties with cancer and heart patients have failed to show any obvious beneficial effects.

Stress, in contrast, struck me from the outset as a deeply unwieldy concept that, notwithstanding its apparent hard-nosed roots in the laboratory, had come to stand in for virtually every class of human unease, distress, and malaise—to the detriment of all involved. One anecdote in particular, which I heard in the late 1990s, clarified the issues for me here. Robert Rose, a psychiatrist and onetime major player in the field of stress research, told me how, in the 1970s, he had been part of a team commissioned by the Federal Aviation Administration to investigate whether (as their union leaders insisted) U.S. air traffic controllers were working under conditions of unbearable stress. A strategy was developed for sorting this situation out, and determining how bad things really were: the amount of "stress" would be quantified (e.g., by counting the numbers of planes controllers had to manage at any one time); and these quantified levels of stress would then be related to quantified measures of malfunctioning (e.g., depression) and maladaptive physiology (e.g., heightened cortisol levels).

As tensions between the FAA, the unions, and the workers increased, however, things began to go wrong. As Rose recalled: "I was focused on the *boom-boom-boom*, the most precise ways of measuring," but the data were not adding up. Having talked with the workers, he had no doubt they were ailing on the job. Yet the cortisol counts (for example) were not validating their suffering in the ways the unions had hoped. By the mid-1970s, Rose and his colleague were thus forced to conclude that "there was no evidence of a stress level effect among air traffic controllers."[19]

Then, the night before the first air traffic controller strike in U.S. history, Rose recalled how he "argued with the president of the union on [the television show] *Nightline*." The air traffic controller crisis wasn't about cortisol levels, high blood pressure, and a shorter work week. "I said to him," Rose recalled: " 'Bob, you and I both *know* that the *real* issue for the controllers is their perception that no one gives a damn, and no one really cares—that they are not being supported.' " It was a bombshell of a reproach—not only to the union managers, but to the entire premise of all the work Rose had done attempting to clarify what was wrong with these men. As he put it to me so many years later, it seemed that actually "cortisol wasn't truth, and stress wasn't just inside the individual." The realities of betrayal, interpersonal alienation, and power politics were all beginning instead to emerge as necessary explanatory categories for making sense of the data. But, he told me, "The agency didn't want to hear about that. They said, 'we hired you to study stress, not that—that's not relevant to the data you collected.' "[20] In the end, therefore, what Rose could blurt out as a self-evident truth in the heat of debate on *Nightline* became impossible for him to articulate in the frame of his work as a stress scientist. Everyone lost.

So, I listened to my colleagues, I made my appraisals, I participated in their discussions, and I tried to make contributions to some of their projects. In the end, however, I always came back to the stories. They seemed key to me, first because they helped me better understand how to relate questions about the *truth* of mind-body medicine to questions about the *meaning* of mind-body medicine, and in this way make better sense of the high-stakes, fraught, boundary-crossing nature of this community in which I had become a minor player. The stories also seemed key to me because they helped me better understand the ways in which my science colleagues both did and did not truly engage with the challenges I felt were raised by mind-body phenomena for broader efforts to talk coherently about what it means to be human.

Let me explain. Throughout this book, I have emphasized the fact that mind-body medicine is a deeply storied world. I have reviewed the historical emergence of the narrative templates that have given rise to the major classes of living stories we see around us today. I have explored the ways in which these different kinds of stories both claim the authority of science and medicine (even if sometimes of an alternative sort) and function as carriers of moral, religious, and existential levels of meaning. Repeatedly, I have called attention to the ways in which variants of specific narrative traditions are told, lived, and experienced by scientists, doctors, alternative healers, patients, cultural critics, and ordinary people. Quite literally, these stories belong to all of us.

One big question, though, remains: why is mind-body medicine such a distinctively storied world? It is true, of course, that any branch of medical and scientific inquiry that tries to make sense of phenomena that change over time is likely to organize its findings in the form of narratives. As historian Misia Landau noted: "The growth of a plant, the progress of a disease, the formation of a beach, the evolution of an organism—any set of events that can be arranged in a sequence and related can also be narrated."[21] The stories of mind-body medicine, however, are much more than just a sequential arrangement of observed events. Structurally, they are different from other kinds of scientific and medical stories because their main task, as narratives, is to knit together domains of experience that we struggle otherwise to relate: the medical and the moral, the biological and the biographical, the natural and the cultural. A story about mind-body processes can begin in the realm of human affairs and morality ("His boss was forever screaming at his employees") and end in the realm of pathophysiology ("One day, he suddenly keeled over with a heart attack"). When we tell or hear such a story, most of us experience a sense of narrative closure—worlds have been brought together and we understand something, even if we are wholly unable to say anything coherent about the specific causal processes by which an unpleasant managerial style turns into an infarction. Stories in this sense allow everyone—scientists, patients, the rest of us—to recognize and speak about the reality of mind-body effects, but to do so in ways that do not require us to confront head-on the age-old dualisms of our culture that we know are wrong, but do not quite know how to fix.

If this is right, then it suggests that we are likely to continue for a long time to use stories across all levels of the culture of mind-body medicine as devices for bridging the lacunae in our thinking. And that is not all. This book has shown that the stories of mind-body medicine do not merely describe experience and behaviors that are given in the world; they also help create behaviors and experiences that had not previously been there. Given this, it seems clear to me that the future of mind-body medicine should lie in its seeking, not finally to escape from its stories, but to embrace them as part of its map and part of its territory alike—inextricably part of, and fundamental to, what it is all about.

Notes

Introduction

1. Steven Ringel, "Patients Like Linda," *Journal of the American Medical Association* 290 (2) (July 9, 2003): 165–66.
2. For more on this, see also Kathryn Montgomery Hunter, *Doctors' Stories* (Princeton, N.J.: Princeton University Press, 1991).
3. Arthur W. Frank, *The Wounded Storyteller* (Chicago: University of Chicago Press, 1995), p. 53.
4. Kathy Quinn Thomas, "The Mind-Body Connection: Granny Was Right, After All," *Rochester Review* 59, no. 3 (1997), available online at www.rochester.edu/pr/Review/V59N3/feature2.html.
5. Timothy J. Reiss, "Denying the Body? Memory and the Dilemmas of History in Descartes," *Journal of the History of Ideas* 57, no. 4 (1996): 587–607, cited p. 592. For a viewpoint that emphasizes Descartes's usefulness as a foil for the makers of the first psychosomatic medicine, see Theodore M. Brown, "Descartes, Dualism, and Psychosomatic Medicine," in *The Anatomy of Madness: Essays in the History of Psychiatry*, vol. 1, ed. W. F. Bynum, Roy Porter, and Michael Shepherd (New York: Tavistock, 1985), pp. 40–62.
6. Cf. here Cheryl Mattingly, *Healing Dramas and Clinical Plots* (New York: Cambridge University Press, 1998).
7. "All the world is not, of course, a stage, but the crucial ways in which it isn't are not easy to specify." See Erving Goffman, *The Presentation of Self in Everyday Life*. (Garden City, N.Y.: Doubleday, 1959), p. 72.
8. Keith Hoskin, "The 'Awful Idea of Accountability': Inscribing People into the

Measurement of Objects," in *Accountability: Power, Ethos and the Technologies of Managing*, ed. Rolland Munro and Jan Mouritsen (London: International Thomson Business Press,1996), pp. 265–82.

9. See Ian Hacking, "The Looping Effect of Human Kinds," in *Causal Cognition: A Multidisciplinary Debate*, ed. Dan Sperber, David Premack, and Ann James Premack (New York: Oxford University Press, 1995), pp. 351–83, cited p. 351.

10. Richard Katz, *Boiling Energy* (Cambridge, Mass.: Harvard University Press, 1982).

11. Margaret Lock, *Encounters with Aging* (Berkley: University of California Press, 1993); Margaret Lock, "Menopause: Lessons from Anthropology," *Psychosomatic Medicine* 60 (1998): 410–19. Other researchers have linked the low incidence of hot flashes in Japan and other Asian countries to high soybean consumption, which is believed to stimulate estrogen production, but more recent studies have failed to confirm the soy explanation. See Lynnette Leidy Sievert, Lynn Morrison, Daniel E. Brown, and Angela M. Reza, "Vasomotor Symptoms among Japanese-American and European-American Women Living in Hilo, Hawaii," *Menopause* 14, no. 2 (2007): 261–69. The authors write: "Japanese-American women reported a higher intake of soy, but soy intake was not associated with fewer vasomotor symptoms."

12. See Caroline Walker Bynum, *Fragmentation and Redemption: Essays on Gender and the Human Body in Medieval Religion* (New York: Zone Books, 1991); Barbara Duden, *The Woman Beneath the Skin* (Cambridge, Mass.: Harvard University Press, 1991).

13. The evocative term "mindful body" was used initially by Nancy Scheper-Hughes and Margaret Lock, "The Mindful Body: A Prolegomenon to Future Work in Medical Anthropology," *Medical Anthropology Quarterly* New Series 1, no. 1 (1987): 6–41.

14. Edward Shorter's *From Paralysis to Fatigue* moves in this direction. Shorter argues that only certain symptoms are legitimate indicators of illness and that, in this sense, societies have "symptom pools" from which patients must choose to be recognized as legitimate sufferers. The pool is powerfully shaped by cultural and (particularly in modern Western society) medical expectations. See Edward Shorter, *From Paralysis to Fatigue* (New York: Free Press, 1992).

15. Taken from a poem written by a fan of the princess and her companion, posted online at a memorializing site called "Modern Day Romeo & Juliet": www.angelfire.com/sys/popup_source.shtml?Category=;. (This link is now inactive.)

16. There are multiple variations on this theme, including ones that throw in twists of plot involving conspiracy and assassination. Two of many examples include Edith M. Lederer, "In the last weeks of her life, Diana finally found happiness," *NewsTimes: International News* (Danbury, Conn.), September 5, 1997, posted online at www.newstimes.com/archive97/sep0597/inb.htm); and Amy Brooks, "Diana, Dodi depict star-cross lovers of 90s," (University of Tennessee, Knoxville) *Daily Beacon*, February 11, 1998, posted online at http://beacon-www.asa.utk.edu/issues/v77/n20/brooks.20v.html. (Both of these links are now inactive.)

17. I am aware that some analysts of narrative and myth, influenced by Jung and others, insist that human beings are capable of narrating only a handful of basic plots:

"rags to riches," "overcoming the monster," "the quest," "voyage and return," "comedy," "tragedy," and "rebirth" are seven recently proposed by British author Christopher Booker. See Christopher Booker, *The Seven Basic Plots* (London: Continuum, 2004). It may or may not be that the stories of American mind-body medicine could be mapped onto such allegedly more fundamental plots, but I have not been particularly interested to attempt an exercise of this sort.

18. The classic work here is that of sociologist Norbert Elias, whose history of the emerging rules of etiquette in modern Europe—of what he called "the civilizing process"—has been both enormously influential and controversial. Elias's work here was first published in 1939 but achieved widespread attention only in the 1960s, when it was republished and translated into English. See Norbert Elias, *The History of Manners* (New York: Pantheon, 1982); and *Power and Civility* (New York: Pantheon, 1982).

Chapter One: The Power of Suggestion

1. Bruno Klopfer, "Psychological Variables in Human Cancer," *Journal of Projective Techniques* 31, no. 4 (December 1957): 331–40.
2. Klopfer obtained a Rorschach record of the patient before his final relapse and determined that it showed Mr. Wright to have a "floating ego" personality style without any "deep-rooted personality center." This, he said, had made Mr. Wright profoundly suggestible to the newspaper reports and the promises of his doctors but had left him without any energetic capacity to resist breakdown and catastrophe when (as Klopfer put it) "the power of suggestion had expired." Klopfer, "Psychological Variables," p. 339.
3. See Erika Bourguignon, "World Distribution and Patterns of Possession States," *Trance and Possession States* (Montreal: R. M. Bucke Memorial Society, 1968), pp. 3–34. See also Bourguignon, *Possession* (San Francisco: Chandler & Sharp, 1976).
4. For a sampling of this perspective, see Tara L. AvRuskin, "Neurophysiology and the Curative Possession Trance: The Chinese Case," *Medical Anthropology Quarterly* 2, no. 3 (1988): 286–302; Peggy Ann Wright, "The Nature of the Shamanistic State of Consciousness: A Review," *Journal of Psychoactive Drugs* 21, no. 1 (January–March 1989): 25–33; Steven Kane, "Holiness Ritual Fire Handling: Ethnographic and Psychophysiological Considerations," *Ethos* 10, no. 4 (Winter 1982): 369–84; Wolfgang Jilek, "Altered States of Consciousness in Northern American Indian Ceremonies," *Ethos* 10, no. 4 (Winter 1982): 326–42. Also see Michael Winkelman, *Shamanism* (Westport, Conn.: Bergin & Garvey, 2000).
5. P. S. Alexander, "The Demonology of the Dead Sea Scrolls," *The Dead Sea Scrolls after Fifty Years*, vol. 2, ed. Peter W. Flint and James C. VanderKam (Leiden: E. J. Brill, 1999), pp. 331–53. See also Samuel S. Kottek, "Demons and Diseases in Bible and Talmud," in *Illness and Healing in Ancient Times* (Haifa, Israel: Reuben and Edith Hecht Museum, 1996), pp. 32–38.
6. Mark 9:14–27. There are some twenty-six references in total to the casting out of devils in the New Testament. For more, see Carl A. Wickland, *30 Years Among the Dead* (1924; London: Newcastle Publishing, 1974).
7. Mark Bancroft, *The History & Psychology of Spirit Possession & Exorcism* (EnSpire

Press, 1998), available online at www.enspirepress.com/writings_on_consciousness/spirit_possession_exorcism/spirit_possession_exorcism.html.

8. Keith Thomas, *Religion and the Decline of Magic* (London: Weidenfeld & Nicolson, 1971), p. 488.
9. In 1999, the Vatican convened a panel of experts to bring the thinking of the Catholic Church on Satan and demonic possession into the modern era. Monsignor Corrado Balducci, the Vatican's chief exorcist, explained that the Church has to "be more careful in distinguishing between possession by evil spirits and what are more commonly called psychiatric disturbances. We are changing the rules for the millennium as part of the continuing process of liturgical reform." According to Vatican officials, priests confronting cases of potential possession would now be encouraged not to refer to the Prince of Darkness, the Accursed Dragon, the Foul Spirit, the Satanic Power, or the Master of Deceit. Instead the formulas would now refer to "the cause of evil." Richard Owen, "Satan Gets a Facelift," *The Australian*, January 26, 1999; available online, for a fee, at www.theaustralian.com.au.
10. Lana Condie, "The Practice of Exorcism and the Challenge to Clerical Authority," *Access: History* 3, no. 1 (2000): 93–103, cited p. 94.
11. D. P. Walker, *Unclean Spirits* (London: Scolar Press, 1981), pp. 34–35, 38. For the Walker reference I am indebted to Richard Olson, "Spirits, Witches, and Science: Why the Rise of Science Encouraged Belief in the Supernatural in 17th-Century England," *Skeptic* 1, no. 4 (1992): 34–43.
12. D. P. Walker, *Unclean Spirits*, p. 12.
13. Thomas Hobbes, *Leviathan: Or the Matter, Form, and Power of a Commonwealth, Ecclesiastical and Civil*. 5th ed. (London: George Routledge and Sons, 1885), p. 311.
14. The first scholar to tell the story of Gassner and Mesmer in any detail was the French Canadian psychiatrist and historian Henri Ellenberger in his still unsurpassed exploration of the multiple sources that created dynamic psychiatry, *The Discovery of the Unconscious* (New York: Basic Books, 1970).
15. Cited in Vincent Buranelli, *The Wizard from Vienna* (London: Peter Owen, 1975), p. 24.
16. For more on Gassner's healings, see Francis J. Schaefer, "Johann Joseph Gassner," in *The Catholic Encyclopedia*, vol. 6 (New York: Robert-Appleton, 1909), available online at www.newadvent.org/cathen/06392b.htm, © 1999.
17. For more on Mesmer's biography—some details of which are in dispute—the more reliable works include Ellenberger, *Discovery of the Unconscious*; Robert Darnton, *Mesmerism and the End of the Enlightenment in France* (Cambridge, Mass.: Harvard University Press, 1968); Buranelli, *Wizard from Vienna*; and Adam Crabtree, *From Mesmer to Freud* (New Haven, Conn.: Yale University Press, 1993.)
18. Ellenberger, *Discovery of the Unconscious*, p. 57.
19. In some instances, the treatment was modified to enable groups of people to be magnetized at the same time. In this variation, patients were made to grasp iron rods that were connected at right angles to a large tub—the famous mesmeric *baquet*. This tub was filled with water that Mesmer had treated in advance with

his own animal magnetism and to which iron filings and powdered glass had been added to augment the magnetic effect. Patients were "encouraged to [further] augment the magnetic fluid by holding hands, thus creating a circuit." Crabtree, *From Mesmer to Freud*, pp. 13–14.

20. Of this genre of literature, Charles Mackay's extended discussion of "the magnetizers" in his 1841 *Extraordinary Popular Delusions and the Madness of Crowds* is perhaps unsurpassed both for its biting prose and its erudition. This book has gone through a series of editions, and is currently in print in a reissued paperback (New York: Crown, 1995). It is also available online at www.econlib.org/library/mackay/macEx7.html.
21. Darnton, *Mesmerism*, pp. 10–11.
22. Buranelli, *Wizard from Vienna*, pp. 110–11.
23. Letter from Jane Carlyle, December 13, 1847, reprinted in Lee Jackson, "The Cat's Meat Shop" [Lee Jackson's blog—a Victorian rag-bag]; available online at catsmeatshop.blogspot.com.
24. The story of Mesmer's failed treatment of Maria Theresa Paradis has been told, from varying perspectives, in every reasonably comprehensive treatment of his life. Buranelli, in his *Wizard from Vienna*, is confident that the charges of impropriety on Mesmer's part were lacking in any foundation (though he suggests that the young Maria Theresa may well have been infatuated with her charismatic doctor). He is one of several who believe that the real issue for those involved in this affair was that a true healing for Maria Theresa would have spelled the end of her fame as a blind pianist and the probable loss of the disability pension set up for her by her patron and namesake, the Empress Maria Theresa.
25. Cited in Darnton, *Mesmerism*, p. 59.
26. Mackay, *Extraordinary Popular Delusions*.
27. Both commissions actually investigated not the practice of Mesmer himself but the mesmeric practice of an ardent follower, the distinguished French physician Charles d'Eslon. D'Eslon had been told to renounce the doctrine of animal magnetism or be expelled from the Faculty of Medicine. He not only refused but declared himself convinced that animal magnetism represented a great scientific and medical discovery. He was therefore ready, he said, to submit his practice to evaluation of whatever sort was deemed appropriate. Mesmer, meanwhile, had previously left Paris for the countryside, having been snubbed in his efforts to secure royal patronage, and offended by the whole proceedings. Though repeatedly invited to return, he chose to stay away during the entire investigation, claiming ill health.
28. After investigating mesmeric treatments for some four months, the members of this strictly medical commission—with one significant dissenting voice, that of the botanist Antoine-Laurent de Jussieu—professed themselves disappointed. They had witnessed no cases in which clearly identifiable disorders had been successfully cured. See *Rapport des Commissaires de la Société Royale de Médecine*, 1784, reprinted in full in Alexandre Bertrand's *Du Magnétisme Animal en France* (Paris: J. B. Baillière, 1826), pp. 482–510. Jussieu's dissenting report was published as *Rapport de l'un des commissaries charges par le roi de l'éxamen du magnétisme animal* (Paris: Vve Héerissant, 1784). It is not widely available in libraries

today; however see Bertrand *Du Magnétisme Animal*, pp. 151–206, for a full reprinting.

29. See Ted J. Kaptchuk, "Intentional Ignorance: A History of Blind Assessment and Placebo Controls in Medicine," *Bulletin of the History of Medicine* 72, no. 3 (1998) 389–433.
30. Benjamin Franklin et al., *Testing the Claims of Mesmerism* (an English translation of the 1784 report by Benjamin Franklin on Mesmerism, trans. Charles and Danielle Salas, introduction by Michael Shermer, *Skeptic* 4, no. 3 (1996): 66–83.
31. On the dangers of the imagination in eighteenth-century European thinking, see, for example, John C. Whale, *Imagination Under Pressure 1789–1832: Aesthetics, Politics, Utility* (Cambridge: Cambridge University Press, 2000); Jan Goldstein, "Enthusiasm or Imagination? Eighteenth-Century Smear Words in Comparative National Context," in *Enthusiasm and Enlightenment in Europe, 1650–1850*, ed. Lawrence E. Klein and Anthony J. La Vopa (San Marino, Calif.: Huntington Library 1998); originally published in *Huntington Library Quarterly* 60 (1998): 29–49.
32. Isabelle Stengers, "The doctor and the charlatan," *Cultural Studies Review* 9 (2) (November 2003): 11–36, cited p. 16. This piece was originally published as "Le Médecin et Le Charlatan," in Tobie Nathan and Isabelle Stengers, *Médecins et Sorciers* (Paris: Le Plessis-Robinson: Synthélabo, 1995).
33. There were some renewed efforts, beginning in the 1830s, to gain a new medical hearing for mesmerism, most notably perhaps by John Elliotson, professor of clinical medicine at University College London. The efforts largely backfired. Elliotson was forced eventually to resign from his position in the face of increasingly vitriolic criticisms of some of his public demonstrations. One area where mesmerism came close to gaining a footing in medicine was as a form of anesthesia during surgery (there were reputable reports of operations carried out on patients in a state of mesmerism without pain and with a greatly reduced death rate). However, these developments were short-circuited in the late 1840s by the triumph in clinical practice of chemical forms of surgical anesthesia like ether. Alison Winter has argued that the alacrity with which chemical anesthesia was accepted by the orthodox medical profession had almost as much to do with its hostility toward mesmerism as it did with its hostility toward pain. See Alison Winter, "Mesmerism and the Introduction of Surgical Anesthesia to Victorian England," *Engineering & Science* 2 (1998): 20–37.
34. Jean Bailly, 1784, in Bertrand, *Du Magnétisme Animal*, pp. 73, 74. The translation from the French here is mine.
35. "Rapport Secret sur le Mésmerisme, ou Magnétisme Animal" (1784), compiled by J. S. Bailly, in Bertrand, *Du Magnétisme Animal*, pp. 511–16. This report was recently translated and published as J. S. Bailly et al., "Secret report on mesmerism, or animal magnetism," *International Journal of Clinical and Experimental Hypnosis* 50, no. 4 (October 2002): 364–68.
36. See, e.g., Peter Linebaugh, *The London Hanged: Crime and Civil Society in the Eighteenth Century* (London: Verso, 2003); Robert W. Malcolmson, *Popular Recreations in English Society, 1700–1850* (New York: Cambridge University Press, 1973).

37. Frank Podmore, *From Mesmer to Christian Science* (New Hyde Park, N.Y.: University Books, 1963), p. 71.
38. Marquis de Puységur, *Mémoires pour server à l'histoire et à l'établissement du magnétisme animal*, 2d ed. (Paris: Cellot, 1809), p. 103.
39. Cited in Podmore, *From Mesmer to Christian Science*, p. 77.
40. See Nicholas Spanos and Jack Gottleib, "Demonic Possession, Mesmerism, and Hysteria: A Social Psychological Perspective on their Historical Interrelation," *Journal of Abnormal Psychology* 88 (1979): 527–46. Spanos has also done research suggesting that, while some symptoms of possession—convulsions, amnesias, increased strength, insensitivity to pain—were common across different sites in Europe and the American colonies, only in Catholic countries like France (where magnetic somnambulism first emerged) did one see situations in which demons spoke through possessed persons as independent personalities. This difference, Spanos suggested, was due to the fact that the Catholic ritual of exorcism involved a process of interrogating and subduing the demon directly. Protestant countries eschewed exorcism as a kind of collaboration with the demonic and instead relied on prayer and fasting to heal possessed persons. Nicholas P. Spanos, "Multiple Identity Enactments: A Social Psychological Perspective"; paper presented October 1992 at the annual meeting of the Society for Clinical and Experimental Hypnosis, Arlington, Va., available online at www.hypnosis-research.org/hypnosis.
41. A good introduction to the complexities and diversities of romanticism is *Romanticism in National Context*, ed. Roy Porter and Mikulas Teich (New York: Cambridge University Press, 1988).
42. See John O. Lyons, *The Invention of the Self* (Carbondale: Southern Illinois University Press, 1978); Gerald N. Izenberg, *Impossible Individuality: Romanticism, Revolution, and the Origins of Modern Selfhood, 1787–1802* (Princeton, N.J.: Princeton University Press, 1992).
43. Anonymous, "New Uses of Mesmerism—Caution to Husbands." The article was republished in the Ohio newspaper *The Experiment* on July 24, 1844, which is where I accessed it.
44. Indeed he had been largely anticipated in his way of thinking by a number of the magnetizers from whom he tried hard to distinguish himself. James Braid, *Neurypnology, or, the Rationale of Nervous Sleep, Considered in Relation with Animal Magnetism* (1843; London: Churchill, 1848).
45. "The first exhibition of the kind I ever had an opportunity of attending, was one of M. Lafontaine's conversazione, on the 13th November, 1841. That night I saw nothing to diminish, but rather to confirm, my previous prejudices. At the next conversazione, six nights afterwards, *one* fact, the inability of a patient to *open his eyelids*, arrested my attention. I considered that to be a *real phenomenon*, and was anxious to discover the physiological cause of it. Next night, I watched this case when again operated on, with intense interest, and before the termination of the experiment, felt assured I had discovered its cause, but considered it prudent not to announce my opinion publicly, until I had had an opportunity of testing its accuracy, by experiments and observation in private." Braid, *Neurypnology*, p. 16.
46. Braid, *Neurypnology*, p. 4.

47. J. M. Charcot, "Sur les divers états nerveux déterminés par l'hypnotisation chez les hystériques," *Comptes Rendues de l'Academie des Sciences* 94 (1882): 403–05.
48. Pierre Janet, *Psychological Healing*, 2 vols. (New York: Macmillan, 1925). Originally published as *Les médications psychologiques* (Paris: Félix Alcan, 1919).
49. Cited in Georges Didi-Huberman, *Invention of Hysteria: Charcot and the Photographic Iconography of the Salpêtrière* (Cambridge, Mass.: MIT Press 2003), p. 76.
50. Cited in Jan Goldstein, "The Hysteria Diagnosis and the Politics of Anticlericalism in Late Nineteenth-Century France," *Journal of Modern History* 54 (June 1982): 209–39, cited p. 215.
51. This claim can be defended, even as the story of Charcot's introduction to hypnosis also involved considerable serendipity and a more complex relationship to the allegedly taboo concepts of animal magnetism than is widely appreciated. I have previously pursued the larger story here in two articles: "Metals and Magnets in Medicine: Hysteria, Hypnosis, and Medical Culture in fin-de-siècle Paris," Psychological Medicine 18, no. 1 (February 1988): 21–38; and "Hysteria, Hypnosis, and the Lure of the Invisible: The Rise of Neo-Mesmerism in fin-de-siècle French Psychiatry," in *The Anatomy of Madness: Essays in the History of Psychiatry,* vol. 3, ed. W. F. Bynum, Roy Porter, and Michael Shepherd (New York: Tavistock, 1989).
52. A-. A. Liébeault, *Du sommeil et des états analogues, considérés surtout au point de vue de l'action du morale sur le physique* (Nancy: Nicolas Grosjean, 1866). This was the book through which Bernheim first became aware of Liébeault's work.
53. Alan Gauld, *A History of Hypnotism* (New York: Cambridge University Press, 1992), p. 324.
54. Hippolyte Bernheim, *De la suggestion dans l'état hypnotique et dans l'état de veille* (Paris: Octave Doin, 1884). A revised and expanded version appeared in 1886 under the title *De la suggestion et de ses applications thérapeutique* (Paris: Octave Doin). This title (in its second revised edition) served as the basis for an English translation, *Suggestive Therapeutics: A Treatise on the Nature and Uses of Hypnotism* (New York: G. P. Putnam's, 1889).
55. Bernheim, *Suggestive Therapeutics*, p. ix.
56. Jean-Martin Charcot, *Charcot the Clinician: The Tuesday Lessons* (New York: Raven Press, 1987), p. 107.
57. Babinski in fact went so far as to propose that medicine abandon the word "hysteria" to describe the syndrome in question, and instead adopt a new term that would solidify a much narrower understanding: "pithiatism." This was a word derived from the Greek word for "persuasion." After some years of deliberation, the Society voted to adopt this neologism. For this, see J. Babinski, "Démembrement de l'hysterie traditionelle: Pithiatisme," *Semaine médicale* 29, no. 1 (January 6, 1909): 3–8.
58. *Trilby* has recently been brought back in print. For more on its larger cultural significance and the "mania" that surrounded its initial publication, see Daniel Pick, *Svengali's Web* (New Haven, Conn.: Yale University Press, 2000), and Emily Jenkins, "*Trilby*: Fads, Photographers, and 'Over-Perfect Feet,' " *Book History* 1, no. 1 (1998): 221–67.
59. "Suggestion," *Encyclopedia Britannica*, 1911.

60. Moriz Benedikt, *Hypnotismus und Suggestion* (Vienna: Breitenstein, 1894), p. 54, cited in David Schur, "Compulsion as Cure: Contrary Voices in Early Freud," *New Literary History* 32, no. 3 (2001): 585–96.
61. Quoted in Gauld, *History of Hypnotism*, pp. 494–95. See also J. Liégeois, *De la suggestion et du somnambulisme dans leurs rapports avec la jurisprudence et la médecine légale* (Paris: Octave Doin, 1889).
62. I am indebted to an article by Martin Plowhouse for a partial transcription of this scene from the film. See Martin Plowhouse, "Missing Time: Alien Abductions and the Secrets of Hypnosis," *Antithesis* (2003), available online at www.english.unimelb.edu.au/antithesis/forum2004/plowman.html.
63. Richard Cabot, "The Use of Truth and Falsehood in Medicine: An Experimental Study," *American Medicine* 5 (1903): 344–49, cited p. 348.
64. Richard Cabot, 1906, cited in R. P. C. Handfield-Jones, "A Bottle of Medicine from the Doctor," *Lancet* 265 (October 17, 1953): 823–25, cited p. 825.
65. See R. P. C. Handfield-Jones, "A Bottle of Medicine from the Doctor"; also S. White, "Medicine's Humble Humbug: Four Periods in the Understanding of the Placebo," *Pharmacy in History* 27, no. 2 (1985): 51–60. For a history of patent medicine in the United States, see J. H. Young, *Medical Messiahs* (Princeton, N.J.: Princeton University Press, 1970; reprinted, with a new introduction by the author, in 1992).
66. O. H. Perry Pepper, *Old Doc* (Philadelphia: J. P. Lippincott, 1957).
67. O. H. Perry Pepper, "A Note on the Placebo," *American Journal of Pharmacy* 117 (1945): 409–12.
68. H. K. Beecher, "The Powerful Placebo," *Journal of the American Medical Association* 159 (1955): 1602–06, cited p. 1604.
69. Later, the placebo arm in clinical trials would also be understood as a way of filtering out not just the suggestive effects of a dummy drug but other kinds of assumed "noise" in the system. As Sir Arthur Bradford Hill, one of the architects of the modern clinical trial methodology, reviewed matters in 1971:

    > The object [of including placebo controls in clinical trials] is two-fold. On the one hand, we hope to be able to discount any bias in patient or doctor in their judgments of the treatment under study. For example, a new treatment is often given a more favorable judgment than its value actually warrants. The effects of suggestibility, anticipation and so on must thus be allowed for. In addition, the placebo effect provides a vital control for the frequency of spontaneous changes that may take place in the course of a disease and are independent of the treatment under study.
    >
    > In these two ways the placebo aids us to distinguish between (a) the pharmacological effect of the drug and (b) the psychological effects of treatment and the fortuitous changes that can take place in the course of time. (Cited in M. Shepherd, "The Placebo: From Specificity to the Non-specific and Back," *Psychological Medicine* 23 [1993]: 1–10, p. 5.)

70. Walter Robinson, "A Misguided Miracle: The Use of Placebos in Clinical Practice," *Ethics & Behavior* 8, no. 1 (1998): 93–95.

Chapter Two: The Body That Speaks

1. Michael Bury, "Illness Narratives, Fact or Fiction?" *Sociology of Health and Illness* 21, no. 3 (2001): 263–85, cited p. 275.
2. This vision of the psychoanalytic session as a secular confessional is far from original. In particular, a version of the claim is at the heart of Michel Foucault's celebrated analysis of the history of sexuality. See Michel Foucault, *The History of Sexuality: An Introduction,* vol. 1 (New York: Random House, 1978).
3. Sigmund Freud, "An Autobiographical Study" (1925), in *The Freud Reader*, ed. Peter Gay (New York: W. W. Norton, 1995), p. 7.
4. Sigmund Freud, *Letters of Sigmund Freud*, ed. Ernst L. Freud (New York: Basic Books, 1960), p. 185.
5. See, for example, Freud's comments on Charcot's views in his "Preface to the Second German Edition of Bernheim's *Suggestion*," 1896, pp. 77–78, in *The Standard Edition of the Complete Psychological Works of Sigmund Freud*, vol. 1: 1886–1899, *Pre-psychoanalytic Publications and Unpublished Drafts* (New York: Vintage [Hogarth Press], 2000). All further references in the notes to this and other volumes in the *Standard Edition* will be designated using the convention *SE* 1, *SE* 2, etc.
6. Letter to Wilhelm Fliess dated December 28, 1887; cited in "Papers on Hypnotism and Suggestion: Editor's Introduction," *SE* 1, p. 64.
7. Freud, "Preface to the Translation of Bernheim's *Suggestion*," *SE* 1, p. 75.
8. Sigmund Freud, 1892–1893, "A Case of Successful Treatment by Hypnotism: With Some Remarks on the Origin of Hysterical Symptoms through 'Counter-Will,' " *SE* 1, pp. 115–28, cited in David Schur, "Compulsion as Cure: Contrary Voices in Early Freud, *New Literary History* 32, no. 3 (2001): 585–96, p. 592.
9. Ibid., pp. 120–21, cited in Schur, "Compulsion as Cure," p. 595.
10. Doubts about the success of Breuer's treatment were voiced within psychoanalytic circles early on, but the critical evidence did not emerge until the early 1970s when the Canadian historian Henri Ellenberger discovered a copy of Breuer's case notes and related documentation. See Henri Ellenberger, "The Story of Anna O: A Critical Review with New Data," *Journal of the History of Behavioral Sciences* 8 (July 1972): 267–79.
11. For more on the life of Bertha Pappenheim, both before and after her treatment under Breuer, see Dora Edinger, *Bertha Pappenheim* (Highland Park, Ill.: Congregation Solel, 1968); and L. Freeman, *The Story of Anna O.* (New York: Walker, 1972).
12. Josef Breuer and Sigmund Freud, 1895, *Studies on Hysteria*, *SE* 2, ed. James Strachey (New York: Basic Books, 1957), p. 21.
13. Ibid., pp. 34–35.
14. Lillian R. Furst, *Just Talk* (Lexington: University Press of Kentucky, 1999), p. 36.
15. Sigmund Freud, 1896, "The Aetiology of Hysteria," *SE* 3, pp.189–221, cited p. 203.
16. Sigmund Freud, 1905, "Three Essays on the Theory of Sexuality," *SE* 7, pp. 136–248. For a fuller discussion of Freud's initial reasons for asserting the truth of the seduction theory and his reasons for subsequently rejecting it, see

J. G. Schimek, "Fact and Fantasy in the Seduction Theory: A Historical Review," *Journal of the American Psychoanalytic Association* 35 (1987): 937–65.

17. Judith Lewis Herman, *Father-Daughter Incest* (Cambridge, Mass.: Harvard University Press, 1980); Jeffrey M. Masson, *The Assault on Truth* (New York: Farrar, Straus & Giroux, 1984).
18. Pierre Janet, "Autobiography of Pierre Janet," first published in *History of Psychology in Autobiography*, vol. 1, ed. Carl Murchison (New York: Russell & Russell, 1961), pp. 123–33). Republished by permission of Clark University Press, Worcester, Mass., © 1930 Clark University Press. Available online at http://psychclassics.yorku.ca/Janet/murchison.htm.
19. For more on this, see Michael S. Roth, "Hysterical Remembering," *Modernism/Modernity* 3 (April 1996): 1–30.
20. Ernest Jones, *Sigmund Freud: Life and Work*, vol. 2 (London: Hogarth Press, 1955), p. 112.
21. C. S. Myers, "A Contribution to the Study of Shell Shock: Being an Account of Three Cases of Loss of Memory, Vision, Smell, and Taste, Admitted into the Duchess of Westminster's War Hospital, Le Touquet," *Lancet*, February 13, 1915: 316–20.
22. Cited in Ted Bogacz, "War Neurosis and Cultural Change in England, 1914–22: The Work of the War Office Committee of Enquiry into 'Shell-Shock,' " *Journal of Contemporary History* 24, no. 2 (April 1989): 227–56, cited p. 239.
23. See M. Nonne, "Über erfolgreiche Suggestivbehandlung der hysteriformen Störungen bei Kriegsneurosen," *Zeitschrift für die Gesamte Neurologie und Psychiatrie* 37 (1917): 191–218.
24. Lewis Yealland, *Hysterical Disorders of Warfare* (London: Macmillan, 1918).
25. William H. R. Rivers, "Freud's Psychology of the Unconscious," *Lancet* (June 16, 1917, republished as an appendix in W. H. R. Rivers (1920), "Instinct and the Unconscious: A Contribution to a Biological Theory of the Psycho-neuroses," available online at *Classics in the History of Psychology*, developed by Christopher D. Greene, http://psychclassics.yorku.ca/Rivers/appendix1.htm.
26. W. H. R. Rivers, "The Repression of War Experience," in the *Proceedings of the Royal Society of Medicine* 11 (1918): 1–17. Originally delivered as an address before the Section of Psychiatry, Royal Society of Medicine, on December 4, 1917.
27. Sigmund Freud, 1919, "Introduction, Psycho-analysis and the War Neuroses," *SE* 17, p. 210.
28. Trevor Harvey Levere, "Coleridge on Dreaming: Romanticism, Dreams, and the Medical Imagination (review)" *Configurations* 7, no. 2 (1999): 294–95.
29. For more on these general trends in Germany, see Anne Harrington, *Reenchanted Science* (Princeton, N.J.: Princeton University Press, 1996).
30. Cited in Lazslo Antonio Avila, "Georg Groddeck: Originality and Exclusion," *History of Psychiatry* 14, no. 1 (2003): 83–101, cited p. 86.
31. Georg W. Groddeck, *The Book of the It* (New York: International Universities Press, 1979), pp. 100–101; English translation of *Das Buch vom Es: Psychoanalytische Briefe an eine Freundin* (Vienna: Internationaler Psychoanalytischer Verlag, 1923).
32. Freud wrote to Groddeck in 1917 that "[Y]ou urge me to confirm to you officially that you are not a psychoanalyst, that you don't belong to the flock of disciples.

But . . . I must lay claim to you, must insist that you are an analyst of the first order who has grasped the essence of the matter once for all." Letter from Freud to Groddeck, June 5, 1917, in *Letters of Sigmund Freud,* ed. Ernst L. Freud (New York: Basic Books, 1960), p. 316.

33. Groddeck, *The Book of the It*, p. 25.
34. For more on later efforts to contain and deny Groddeck's influence and relevance, see Avila, "Georg Groddeck." As late as 1988, medical anthropologist Arthur Kleinman still found it necessary to warn against the seductions of Groddeck-style psychosomatics: "[F]or all its fascination and promise, [it] leads to the dead end of speculation and an absence of research." Arthur Kleinman, *The Illness Narratives* (New York: Basic Books, 1988), p. 40.
35. In Karen Brecht, "In the Aftermath of Nazi-Germany: Alexander Mitscherlich and Psychoanalysis—Legend and Legacy" *American Imago* 52, no. 3 (1995): 291–312, cited p. 296.
36. In the 1920s, Weizsäcker collaborated with Buber and the Catholic theologian Joseph Wittig on a short-lived interfaith journal called *Die Kreatur*, "The Creature." Each of these men saw himself as an "exile" from three distinct faith traditions (Judaism, Protestantism, and Catholicism), who had as their collective goal to contribute to the effort to establish a new religious sensibility suited to the needs of the modern age.
37. See, e.g., Weizsäcker's "Der Arzt und der Kranke: Stücke einer medizinischen Anthropologie," *Die Kreatur* 1 (1927): 69–86.
38. Viktor von Weizsäcker, "Studien zur Pathogenese: Angina Tonsillaris" 1946, reprinted in *Psychosomatische Medizin: Wege der Forschung*, ed. Peter Hahn (Darmstadt: Wissenschaftliche Buchgesellschaft, 1985), p. 128.
39. Ibid.
40. Viktor von Weizsäcker, 1927, "Über medizinische Anthropologie," in *Gesammelte Schriften in 10 Bänden, Band 5*, ed. Peter Achilles, Dieter Janz, Martin Schrenk, and Carl Friedrich von Weizsäcker (Frankfurt am Main: Surkamp Verlag, 1986), p. 177.
41. Brecht, "In the Aftermath of Nazi-Germany."
42. See, for example, Manfred Pflanz and Thure von Uexkuell, "Guide to Psychosomatic Literature since 1945," *Journal of Psychosomatic Research* 3 (1958): 56–71.
43. The career of Felix Deutsch is worthy of more attention than I am able to give it here. Readers interested in learning more might begin with his classic edited collection *On the Mysterious Leap from the Mind to the Body: A Workshop Study of the Theory of Conversion* (New York: International Universities Press, 1959).
44. See, for example, Matthias H. Goering, "Körperliche Erkrankungen als auswirkungen seelisher Störungen," *Hippokrates* 5 (1935): 485–90; and his 1937 book *Über seelish bedingte echte Organenkrankungen* (Stuttgart: Hippokrates Verlag). For some of the scholarship that has explored the continuing influence of psychoanalysis more generally under the Nazis, see Robert Cocks, *Psychotherapy in the Third Reich: The Göring Institute* (New York: Oxford University Press, 1985); and the controversial but evocative three-volume study by German literary critic Laurence A. Rickels, *Nazi Psychoanalysis* (Minneapolis: University of Minnesota Press, 2002).
45. According to Alexander Mitscherlich, as told to Hannah Decker. See Hannah S.

Decker, "The Reception of Psychoanalysis in Germany," *Comparative Studies in Society and History* 24, no. 4 (1982): 589–602.

46. The remarks defending medical euthanasia were made in a lecture in 1933 but were first published in 1935. See Viktor von Weizsäcker, "Arztliche Fragen. Vorlesungen über allgemeine Theorie," republished in *Der Arzt und der Kranke: Stücke einer medizinischen Anthropologie*, vol. 5 of *Gesammelte Schriften*, ed. Peter Achilles, D. Janz, et al. (Frankfurt am Main: Surkamp Verlag, 1986), p. 323.
47. The column was called "Out of the Clinic—for the Clinic" ("Aus der Praxis—für der Praxis") and consisted of clinical cases presented often in the form of a dialogue between the physician and the patient, followed by discussion and analysis.
48. See Alfred Haug, *Die Reichsarbeitgemeinschaft für eine Neue Deutsche Heilkunde (1935–36)* (Husum: Matthiesen Verlag, 1985), and Detlef Bothe, *Neue Deutsche Heilkunde, 1933–1945* (Husum: Matthiesen Verlag, 1991).
49. I discussed Weizsäcker's complex relationship to National Socialism in more detail in the last chapter of my book *Reenchanted Science* in the section entitled "What does this have to do with my father's work?"
50. See P. Riedesser and A. Verderber, *"Maschinengewehre hinter der Front." Zur Geschichte der deutschen Militärpsychiatrie* (Frankfurt am Main: Fischer Taschenbuch, 1996); Ben Shephard, *A War of Nerves* (Cambridge, Mass.: Harvard University Press, 2001), pp. 209–312.
51. H. Flanders Dunbar, *Emotions and Bodily Change: A Survey of Literature on Psychosomatic Interrelationships* (New York: Columbia University Press, 1935).
52. Dunbar's life, work, and remarkable interdisciplinary background—she was trained as a Dante scholar, a theologian, a physician, and a psychoanalyst—have been the study of extensive research by Robert C. Powell. See Robert C. Powell, "Healing and Wholeness: Helen Flanders Dunbar (1902–1959) and an Extra-Medical Origin of the American Psychosomatic Movement, 1906–1936," Ph.D. diss., Duke University, 1974; "Helen Flanders Dunbar (1902–1959) and a Holistic Approach to Psychosomatic Problems. I. The Rise and Fall of a Medical Philosophy," *Psychiatric Quarterly* 49, no. 2 (1977): 133–52; "Helen Flanders Dunbar (1902–1959) and a Holistic Approach to Psychosomatic Problems. II. The Role of Dunbar's Nonmedical Background, *Psychiatric Quarterly* 50, no. 2 (1978): 144–57; and "Emotionally, Soulfully, Spiritually 'Free to Think and Act': The Helen Flanders Dunbar (1902–59) Memorial Lecture on Psychosomatic Medicine and Pastoral Care," *Journal of Religion and Health* 40, no. 1 (2001): 97–114.
53. H. Flanders Dunbar, "Character and Symptom Formation: Some Preliminary Notes with Special Reference to Patients with Hypertension, Rheumatic, and Coronary Disease," *Psychoanalytic Quarterly* 8 (1939): 18–47, cited pp. 21–22.
54. "A person who is accident prone would probably do better in a commando or a paratroop unit or in some other more or less individualized and adventurous assignment," Dunbar suggested, "just as persons who are potential sufferers from cardiovascular disease, if they follow the coronary or anginal pattern, will do better and maintain health longer if given recognition and authority." H. Flanders Dunbar, "Medical Aspects of Accidents and Mistakes in the Industrial Army and in the Armed Forces," *War and Medicine* 4 (1943): 161–75, cited p. 171. See also H. Flanders Dunbar and Leon Brody, "Basic Aspects and Applications of the

Psychology of Safety" (New York: Center for Safety Education, New York University, 1959).

55. Stanley L. Engelbardt, "Accidents. Why They Happen—How You Can Avoid Them," *Family Weekly*, August 17, 1969, p. 5.

56. W. H. Auden, *Collected Poems*, ed. Edward Mendelsohn (New York: Vintage Books, 1991), p. 158.

57. For a sampling of the literature on the cancer-prone personality from this period, see C. L. Bacon, R. Renneker, and M. Cutler, "A Psychosomatic Survey of Cancer of the Breast," *Psychosomatic Medicine* 14 (1952): 453; L. LeShan and R. E. Worthington, "Personality as a Factor in the Pathogenesis of Cancer," *British Journal of Medical Psychology* 29 (1956): 49–96; L. LeShan and R. E. Worthington, "Some Recurrent Life History Patterns Observed in Patients with Malignant Disease," *Journal of Nervous and Mental Diseases* 124 (1956): 460–65; A. Coppens and M. Metcalfe, "Cancer and Extroversion," *British Medical Journal* 236 (1963): 18–19; S. Fisher and S. E. Cleveland, "Relationship of Body Image to Site of Cancer," *Psychosomatic Medicine* 18 (1956): 304–09; M. Reznikoff, "Psychological Factors in Breast Cancer: A Preliminary Study of Some Personality Trends in Patients with Cancer of the Breast," *Psychosomatic Medicine* 17 (1955): 96–108; J. A. M. Meerlo, "Psychological Implications of Malignant Growth," *British Journal of Medical Psychology* 27 (1954): 210.

58. Cited in Susan Sontag, *Illness as Metaphor, and AIDS and Its Metaphors* (New York: Picador, 2001, p. 22; *Illness as Metaphor* originally published in 1978). This claim was no casual or desperate defense on Mailer's part, but a conviction that he has continued to develop and defend over decades. In the novel *An American Dream,* the protagonist Rojack reflects at one point: "Cancer is the growth of madness denied. In that corpse I saw, madness went down to the blood—leucocytes gorged the liver, the spleen, the enlarged heart and violet-black lungs, dug into the intestines, germinated stench." And in Mailer's 1982 book *Pieces and Pontifications*, he returned to the scene of his 1960 crime, so to speak, and reflected: "In those days, he got to hate 'The Star Spangled Banner' [on television]. It sounded like the first martial strains of that cancer he was convinced was coming on him, and who knows? If he had not stabbed his wife, he might have been dead in a few years himself—our horror of violence is in its unspoken logic." And later in that book, he reflected again: "If a boy beats up an old woman, he may be protecting himself by discharging a rage which would destroy his body if it were left to work on the cells. . . . [He] may be anything from a brute to Raskolnikov. It requires an exquisite sense of context and a subtle gift as a moralist to decide these matters at times." All Mailer quotes cited in Stefan Kanfer, "Adrenaline and Flapdoodle," a review of Norman Mailer's *Pieces and Pontifications* (New York: Little Brown, 1982), *Time*, June 28, 1982, available online at www.time.com/time/magazine/article/0,9171,949524,00.html.

59. On the role played by the Rockefeller Foundation in both funding and shaping the direction of Alexander's research, see Theodore M. Brown, "Alan Gregg and the Rockefeller Foundation's Support of Franz Alexander's Psychosomatic Research," *Bulletin of the History of Medicine* 61, no. 2 (1987): 155–82.

60. Franz Alexander, *Psychosomatic Medicine* (New York: W. W. Norton, 1950). For a more succinct summary of Alexander's approach see Franz Alexander, "Funda-

mental Concepts of Psychosomatic Research," *Psychosomatic Medicine* 5 (1943): 205–10.

61. H. Weiner, M. Thaler, M. F. Reiser, et al., "Etiology of Duodenal Ulcer—Relation of Specific Psychological Characteristics to Rate of Gastric Secretion (Serum Pepsinogen)," *Psychosomatic Medicine* 19 (1957): 1–10.
62. "Adelaide's Lament," from *Guys and Dolls*, by Frank Loesser.
63. These developments have been gracefully analyzed by Theodore M. Brown in his paper "The Rise and Fall of American Psychosomatic Medicine," presented to the New York Academy of Medicine, New York, November 29, 2000. Available online at: www.human-nature.com/free-associations/riseandfall.html.
64. Cited in Brown, "Rise and Fall of American Psychosomatic Medicine."
65. Steven Hyman, "Another One Bites the Dust: An Infectious Origin for Peptic Ulcers," *Harvard Review of Psychiatry* 1, no. 5 (January–February 1994): 294–95.
66. This website is being regularly updated, and so the details of its exact position on asthma may change slightly over time. See http://yosemite.epa.gov/ochp/ochpweb.nsf/homepage.
67. Judith Lewis Herman, *Trauma and Recovery* (1989; New York: Basic Books, 1997), p. 3.
68. Bessel A. Van der Kolk, "The Body Keeps the Score: Approaches to the Psychobiology of Posttraumatic Stress Disorder," in *Traumatic Stress*, ed. Bessel A. Van der Kolk, A. C. McFarlane, and L. Weisaeth (New York: Guilford, 1996), pp. 214–41; Alice Miller, *The Body Never Lies: The Lingering Effect of Cruel Parenting* (New York: W. W. Norton, 2005); Babette Rothschild, *The Body Remembers* (New York: W. W. Norton, 2000).
69. Christiane Northrup, online biography, available at www.webmd/com/Christiane-Northrup.
70. Christane Northrup, *Women's Bodies, Women's Wisdom* (New York: Bantam Books, 1998), p. 363.
71. Sontag, *Illness as Metaphor, and AIDS and Its Metaphors*, p. 3.
72. "Researchers Looking at Link Between Personality, Cancer," *Post* (Frederick, Md.), January 26, 1983.
73. Lydia Temoshok and Henry Dreher, *The Type C Connection* (New York: Random House, 1992), p. 8. For the work leading to Temoshok's own conclusions (she coined the term "Type C" and contrasted it with the "Type A" personality), see Lydia Temoshok, "Biopsychosocial Studies on Cutaneous Malignant Melanoma: Psychosocial Factors Associated with Prognostic Indicators, Progression, Psychophysiology, and Tumor-Host Response," *Social Science and Medicine* 20 (1985): 833–40.
74. Virginia Lee, "Everyday Miracles: An Interview with Dr. Bernie Siegel," Common Ground On Line: Resources for Personal Tranformation, 1995, posted online at www.comngrnd.com/siegel.html. This interview is no longer available on the internet; I am happy to share my saved digital copy of the interview with anyone who is interested.
75. Bernie Siegel, *Love, Medicine, and Miracles* (1986; New York: HarperPerennial, 1990), p. 99.
76. Ibid., p. 94
77. Ibid., p. 67.

78. Ibid., p. 77.
79. Janet Colli, "Cancer and culture: from Koch to Colli," *Surviving! A Cancer Patient Newsletter*, Fall 1995. This online source is no longer available on the internet; I can provide a digital copy to anyone interested.
80. Katherine Russell Rich, *The Red Devil: To Hell with Cancer—And Back* (New York: Crown, 1999), p. 52. For the record, Bernie Siegel has made a point of saying that he does not shave his head to claim some kind of kinship with cancer patients undergoing chemotherapy. As he told an interviewer: "Carl Jung describes the barber who shows up and shaves the hero's head. He goes on to talk about 'tonsure,' which is the ritual shaving of a monk's head as a symbol for uncovering one's spirituality. When I read that, something went on like a light. I knew it's what I had to do. And you have to remember, when I first shaved my head 18 years ago, it was not in style. It was considered rather bizarre. And by the way, it had nothing to do with patients in chemotherapy." Lee, "Everyday Miracles."
81. Andrew Vickers, "Alternative Cancer Cures: 'Unproven' or 'Disproven'?" *CA: A Cancer Journal for Clinicians* 54 (2004): 110–18. The original study discussed by Vickers (of which, to his credit, Bernie Siegel was an author) is G. A. Gellert, R. M. Maxwell, and B. S. Siegel, "Survival of Breast Cancer Patients Receiving Adjunctive Psychosocial Support Therapy: A 10-Year Follow-up Study," *Journal of Clinical Oncology* 11 (1993): 66–69.
82. Linda S. Griffs, "Life Line: Cancer-Prone Personality Shrine—2004," Life-after-treatment support for Breast Cancer Survivors, available online at www.QuestCycles.com, © 2005.

Chapter Three: The Power of Positive Thinking

1. "Ralph Waldo Emerson, "An Address Delivered Before the Senior Class in Divinity College, Cambridge from 1838," in Ralph Waldo Emerson, *Selected Essays*, ed. Larzer Ziff (New York: Penguin, 1985), pp. 107–28, cited p. 125.
2. Gina Kolata, "Scientific Myths That Are Too Good to Die," *New York Times*, December 6, 1998.
3. Luke 8:48 and 17:19; Mark 10:52; Matt. 9:29–30.
4. For more on Lourdes and its history in this time, see Ruth Harris, *Lourdes* (New York: Viking, 1999).
5. The cautious 1999 report on this case, issued by Patrick Thellier, doctor-in-charge of the Medical Bureau, after a majority vote on November 14, 1998, had the following to say about this case: "It is possible to conclude with a good margin of probability that Mr. Bély suffered an organic infection of the type Multiple Sclerosis in a severe and advanced stage of which the sudden cure during a pilgrimage to Lourdes corresponds with an unusual and inexplicable fact to all the knowledge of science. It is impossible to say any more today in medical science. It is however to the religious authorities to make a pronouncement on the other dimensions of this cure." See "Science Surrenders to the Inexplicable: Jean-Pierre Bély Was Instantly Cured of Multiple Sclerosis," ZENIT news agency, February 11, 1999, available online at www.zenit.org/english/archive/9902/ZE990211/html.

6. Georges Bertrin, "Notre-Dame de Lourdes," in *The Catholic Encyclopedia*, vol. 9 (New York: Robert Appleton, 1910), available online at www.newadvent.org/cathen/09389b.htm, © 2007 by Kevin Knight.
7. Cited in Harris, *Lourdes*, p. 350.
8. Ibid., p. 351.
9. Jean-Martin Charcot, "La foi qui guérit," *Revue Hebdomadaire* 5 (1892): 112–32; published in English as "The Faith Cure," *New Review* 11 (1892): 244–62.
10. Kathleen Ann Comfort, "Divine Images of Hysteria in Emile Zola's *Lourdes*," *Nineteenth Century French Studies* 30, nos. 3 and 4 (2002): 330–46.
11. For an overview, see Jason Szabo, "Seeing Is Believing? The Form and Substance of French Medical Debates over Lourdes," *Bulletin of the History of Medicine* 76, no. 2 (2002): 199–230.
12. H. Flanders Dunbar, "What Happens at Lourdes? Psychic Forces in Health and Disease," *Forum* 91 (1934): 226–31, cited in R. C. Powell, "Mrs. Ethel Phelps Stokes Hoyt (1877–1952) and the Joint Committee on Religion and Medicine (1923–1936): A Brief Sketch," *Journal of Pastoral Care*, no. 2 (June 29, 1975): 99–105.
13. Walter B. Cannon, "The Role of Emotions in Disease," *Annals of Internal Medicine* 9, no. 11 (May 1936): 1453–65, cited pp. 1455–56.
14. Henry Adams, *The Education of Henry Adams*, 1907, cited in John T. McGreevy, "Bronx Miracle," *American Quarterly* 52, no. 3 (2000): 405–43.
15. Alongside "mind cure," the evangelical Protestant tradition in late-nineteenth-century America also had its own analogue to the Catholic "faith cure"—healing by faith—that attracted some modest, if mostly skeptical, medical interest. See Raymond J. Cunningham, "From Holiness to Healing: The Faith Cure in America, 1872–1892," *Church History* 43, no. 4. (December 1974): 499–513.
16. See, for instance, Robert C. Fuller, *Alternative Medicine and American Religious Life* (New York: Oxford University Press, 1989), and Norman Gevitz, ed., *Other Healers: Unorthodox Medicine in America* (Baltimore: Johns Hopkins University Press, 1988).
17. The alternative medicine of this period also included such bodily based healing systems as Thomsonism—a gentle healing method based on ideas about vital energies—homeopathy, and therapeutic mesmerism. See John Harley Warner, "Medical Sectarianism, Therapeutic Conflict, and the Shaping of Orthodox Professional Identity in Antebellum American Medicine," in *Medical Fringe and Medical Orthodoxy, 1750–1850*, ed. W. F. Bynum and Roy Porter (London: Croom Helm, 1987), pp. 234–60.
18. For more on the gendered aspects of the mind-cure movement, see Donald Meyer, *The Positive Thinkers* (New York: Pantheon Books, 1980); Gail Thain Parker, *Mind Cure in New England* (Hanover, N.H.: University Press of New England, 1973); and most recently and specifically, Beryl Satter, *Each Mind a Kingdom* (Berkeley: University of California Press, 1999).
19. Mary Baker Eddy, "Retrospection and Introspection" (Boston: Trustees under the Will of Mary Baker G. Eddy, 1891), available online at www.christianscience.org/Prose%20Works/RetroIntro.html.
20. Matt. 9:2–7.
21. The Metaphysical College no longer functions as a training center but still con-

venes as a one-week class once every three years. Today, people wishing to train as practitioners study instead with a Church-authorized teacher.

22. Rodney Stark, "The Rise and Fall of Christian Science," *Journal of Contemporary Religion* 13, no. 2 (1998): 189–214.
23. In fact, the bylaws do not forbid parents to seek medical care for children. For further discussion, see Rennie B. Schoepflin, *Christian Science on Trial* (Baltimore, Md.: Johns Hopkins University Press, 2003).
24. Report on talk given by Virginia S. Harris, C.S.B., "The Future of Medicine—and the Medicine of the Future," *Christian Science Sentinel*, March 2000. Harris was at that time chairman of the board of directors of the Church. Some dissidents within the Church have apparently criticized the leadership for "misrepresenting Christian Science by identifying it with Mind/Body, New Age, and medical trends." See, on this, the column by Alex Beam, "Who Is Interested in Doctrinal Disputes Inside the Christian Science Church? I Am," *Boston Globe*, November 12, 2002.
25. Marceil DeLacy, Christian Science practitioner and member of the Christian Science Board of Lectureship, March 3, 1999, as part of remarks to a Harvard undergraduate class on the history of mind-body medicine, quoted by permission.
26. *The Quimby Manuscripts* were edited and published in 1921, partly in an effort to resolve the issue of Eddy's indebtedness to Quimby; see P. P. Quimby, *The Quimby Manuscripts*, ed. Horatio W. Dresser (New York: T. Y. Crowell, 1921). For a vast selection of online material on Quimby, including an online edition of Dresser's edited text, see www.ppquimby.com. The controversy over the status of those manuscripts and their role in the history of Christian Science was recently renewed with the publication of a new biography of Mary Baker Eddy by Gillian Gill which took quite a critical position on Quimby. See Gillian Gill, *Mary Baker Eddy,* Radcliffe Biography Series (Reading, Mass.: Perseus Books, 1998). For an indignant review by a New Thought follower, see Deb Whitehouse, "Review of *Mary Baker Eddy* by Gillian Gill," *Journal of the Society for the Study of Metaphysical Religion* 5, no. 1 (Spring 1999): 75–79, available online at http://websyte .com/alan/eddyrevw.htm.
27. Charles Braden, *Spirits in Rebellion* (Dallas, Tex.: Southern Methodist University Press, 1963).
28. Ralph Waldo Trine, *In Tune with the Infinite* (New York: Dodge Publishing, 1919), chapter 2, "Fullness of Life—Bodily Health and Vigor," pp. 82–84.
29. Emma Curtis Hopkins, *Scientific Christian Mental Practice* (Cornwall Bridge, Conn.: High Watch Fellowship, 1958), available online at http:// emmacurtishopkins.wwwhubs.com/scmp.html.
30. "Doctor" Jean Hazzard, *The Mind-Cure Mentor, a Hand-Book of Healing, a Textbook of Treatments, a Compendium of Practical Christian Science*, 1887, quoted in Norman Beasley, *The Cross and the Crown* (New York: Duell, Sloan, and Pearce, 1952), pp. 153–54.
31. William James, "The Religion of Healthy-Mindedness," lectures 4 and 5 from *The Varieties of Religious Experience* (1902; New York: Wayne Proudfoot [Barnes and Noble Classics], 2004), p. 95.
32. Wallace D. Wattles, *The Science of Getting Rich* (Holyoke, Mass.: E. Towne,

1910), chapter 5, "Increasing Life," available online at www.successdoctor.com/wattles/index.htm.

33. Émile Coué, *Self Mastery through Conscious Autosuggestion* (New York: American Library Services, 1922).

34. Roy E. Plotnick, "In Search of Watty Piper: A Brief History of the 'Little Engine' Story: Now Celebrating One Hundred Years of Thinking I Can!" Available online at http://tigger.uic.edu/~plotnick/littleng.htm, © Roy Plotnick 1996, rev. 2006.

35. Norman Vincent Peale, *The Power of Positive Thinking* (New York: Prentice-Hall, 1952), p. 1.

36. For more on the history of the therapeutic turn in postwar American Protestantism, see Joel James Shuman and Keith G. Meador, *Spirituality, Medicine, and the Distortion of Christianity* (New York: Oxford University Press, 2003).

37. Carol V. R. George, *God's Salesman* (New York: Oxford University Press, 1993), p. 90.

38. Braden, *Spirits in Rebellion*, p. 389.

39. Norman Vincent Peale, *Positive Imaging: The Powerful Way to Change Your Life* (Grand Rapids, Mich.: Revell, 1982), pp. 42, 91.

40. Normal Vincent Peale, *Positive Thinking for a Time Like this* (Englewood Cliffs, N. J.: Prentice-Hall, 1975), p. 116 (from a chapter titled "Healthy Thinking and Healthy Feeling").

41. Norman Vincent Peale, *The Amazing Results of Positive Thinking* (Englewood Ciffs, N.J.: Prentice-Hall, 1959), p. 214.

42. "1950s Bestsellers, provided by Cader Books," www.caderbooks.com/best50.html.

43. George, *God's Salesman*, p. 131.

44. Cited in David W. Cloud, "Norman Vincent Peale: Apostle of Self-Esteem," O *Timothy* 11, no. 2 (1994), available online at www.wayoflife.org/fbns/fbns473.html.

45. Ibid.

46. George R. Vaillant, "Tuberculosis: An Historical Analogy to Schizophrenia," *Psychosomatic Medicine* 24, no. 3 (1962): 225–33, cited p. 231.

47. Norman Cousins, "Anatomy of an Illness (as Perceived by the Patient)," *New England Journal of Medicine* 295, no. 26 (December 23, 1976): 1458–63, cited p. 1459.

48. Ibid., p. 1462.

49. Later in life, Cousins would actually accept a position on the board of directors of Ernest Holmes's New Thought church, Religious Science. See John Weldon and Paul Carden, "Ernest Holmes and Religious Science," *News and Research Periodical of the Christian Research Institute* 7, no 1 (1984), available online at http://associate.com/ministry_files/The_Reading_Room/False_Teaching_n_Teachers_3/Religious_Science.shtml.

50. The comment was made by sociologist Florence Ruderman, in a lecture to the New York Academy of Medicine. Cited in Constance Holden, "Cousins' Account of Self-Cure Rapped," Science 214 (November 1981): 892.

51. The classic critique in this time was Ivan Illich's *Limits to Medicine: Medical Nemesis: The Expropriation of Health* (New York: Penguin, 1977).

52. Gay Luce and Erik Peper, "Mind over Body, Mind over Mind," *New York Times,*

September 12, 1971. For a contemporary analysis of the so-called holistic medicine movement of the 1970s, see Sally Guttmacher, "Whole in Body, Mind & Spirit: Holistic Health and the Limits of Medicine," *Hastings Center Report* 9, no. 2 (1979): 15–21. For more information about some of its historical and institutionalized forms, see James Whorton, *Nature Cures: The History of Alternative Medicine in America* (New York: Oxford University Press, 2004).

53. Norman Cousins, *Anatomy of an Illness as Perceived by the Patient* (New York: W. W. Norton, 1979).

54. Cousins, "Anatomy of an Illness (as Perceived by the Patient)," *New England Journal of Medicine*, p. 1463.

55. J. D. Levine, N. C. Gordon, and H. L. Fields, "The Mechanism of Placebo Analgesia," *Lancet* 2, no. 8091 (1978): 654–57.

56. Robert Ader and Nicholas Cohen, "Behaviorally Conditioned Immunosuppression," *Psychosomatic Medicine* 37, no. 4 (July–August 1975): 333–40.

57. See, for an influential example of this kind of thinking, Sissela Bok, "The Ethics of Giving Placebos," *Scientific American* 231 (November 1974): 17–23.

58. O. Carl Simonton, Stephanie Matthews-Simonton, and James Creighton, *Getting Well Again* (Los Angeles: J. P. Tarcher, 1978).

59. S. Greer, T. Morris, and K. W. Pettingale, "Psychological Response to Breast Cancer: Effect on Outcome," *Lancet* 2 (1979): 785–87; K. W. Pettingale, T. Morris, S. Greer, and J. L. Haybittle, "Mental Attitudes to Cancer: An Additional Prognostic Factor," *Lancet* 2 (1985): 750.

60. Patricia Norris, "Belief, Attitudes, and Expectations: The Relationship between Personality and Cancer," excerpted from Garrett Porter and Patricia Norris, *Why Me?* (New York: E. P. Dutton, 1985), available online at www.healthy.net/scr/Article.asp?Id=412.

61. John Lauristen, "The Epidemiology of Fear," *New York Native*," August 1, 1988, reprinted in John Lauristen, *Poison by Prescription* (New York: Asklepios, 1990), available online at www.virusmyth.net/aids/data/jlfear.htm.

62. Sanford I. Cohen, "Voodoo Death, the Stress Response, and AIDS," in *Psychological, Neuropsychiatric, and Substance Abuse Aspects of AIDS*, ed. T. Peter Bridge, Allan F. Mirsky, and Frederick K. Goodwin (New York: Raven Press, 1988), pp. 95–107, cited p. 95.

63. George Solomon and Lydia Temoshok, "A Psychoneuroimmunologic Perspective on AIDS," in *Psychosocial Perspectives on AIDS*, ed. Lydia Temoshok and Andrew Baum (Hillsdale, N.J.: Lawrence Erlbaum, 1990), p. 249.

64. Mary Klaus, *Patriot News* (Harrisburg, Penn.), November 14, 1994. Also worthy of note was a remarkable website called "AIDS, Medicine, and Miracles." Among its offerings was a poem that could have come from a classic New Thought text: "*How Am I? / I don't describe / How I am 'physically' / Because in some ways / It is immaterial / But more so / It is both 'history' / And not / What I want / To give energy to / My aches and pains / My truth / Then truly is / I feel "wonderful" I see love and 'do' / From inside my body /—Out / I don't let anything / Negative / That might happen to me / Interfere / With the reality / Of my joy / I stay truly in / And at peace / As best I'm able / My universe is good / -My caring about me/ is not.*" At the bottom of his poem, the author, Dan Martell, appended the following brief note: "The word's hiv and aids are lower cased to reduce the harm their fear creates." Dan

Martel, originally cited at www.csd.net/~amm/poetry.htm#howami (originally accessed January 4, 2000). This link is no longer live but can be accessed at the Internet Archive (an online library of previously published websites): http://web.archive.org/web/20010308162859/http://www.csd.net/~amm/poetry.htm.

65. Lennore S. Van Ora, "Front-line Report from the Cancer War," *New York Times*, April 17, 1988.

66. Marcia Angell, "Disease as a Reflection of the Psyche," *New England Journal of Medicine* 312, no. 4 (June 13, 1985): 1570–72.

67. M. Watson, J. S. Haviland, S. Greer, J. Davidson, and J. M. Bliss, "Influence of Psychological Response on Survival in Breast Cancer: A Population-based Cohort Study," *Lancet* 354 (October 16, 1999): 1331–36.

68. Kathleen Doheny, "Cancer Myths Abound, Survey Finds," *Health on the Net Foundation*, June 27, 2005, available online at www.lifespan.org/healthnews/2005/6/27/article526502.html.

69. Anne Harrington, ed., *The Placebo Effect* (Cambridge, Mass.: Harvard University Press, 1997); Howard Brody and Daralyn Brody, *The Placebo Response* (New York: Cliff Street Books, 2000); Howard Spiro, *The Power of Hope: A Doctor's Perspective* (New Haven, Conn.: Yale University Press, 1998); Irving Kirsch, *How Expectancies Shape Experience* (Washington, D.C.: American Psychological Association, 1999); Daniel E. Moerman, *Meaning, Medicine, and the "Placebo Effect"* (New York: Cambridge Unversity Press, 2002); Virginia A. Hoffman, Anne Harrington, and Howard Field, "Pain and the Placebo: What We Have Learned," *Perspectives in Biology and Medicine* 48, no. 2 (2005): 248–65.

70. Margaret Talbot, "The Placebo Prescription," *New York Times Magazine*, January 9, 2000.

71. Ibid.

72. Lorette Kuby, *Faith and the Placebo Effect: An Argument for Self Healing* (Novato, Calif.: Origin Press, 2001).

73. Brian Volck, "Faith as 'Wellness Technique'?" Review of *Heal Thyself: Spirituality, Medicine, and the Distortion of Christianity* by Joel James Shuman and Keith G. Meador, in *America (National Catholic Weekly)* 188, no. 13 (April 14, 2002), available online at www.americamagazine.org/BookReview.cfm?articletypeid=31&textID=2919&issueID=430.

74. Asbjorn Hrobjartsson and Peter C. Gotzsche, "Is the Placebo Powerless? An Analysis of Clinical Trials Comparing Placebo with No Treatment," *New England Journal of Medicine* 344, no. 21 (2001): 594–1602; "The Powerful Placebo and the Wizard of Oz," *New England Journal of Medicine* 344 (2001): 1630–32; D. Spiegel, H. Kraemer, R. W. Carlson, et al., "Is the Placebo Powerless?" correspondence, *New England Journal of Medicine* 345 (2001): 1276–79.

75. P. Petrovic et al., "Placebo and Opioid Analgesia: Imaging a Shared Neuronal Network," *Science* 295 (2002): 1737–40.

Chapter Four: Broken by Modern Life

1. Norman Vincent Peale, "The Way to Conquer Stress," *Oshkosh (Wisconsin) Daily Northwestern*, March 16, 1957.

2. "Stress may be worst killer of the modern era," *Press Telegram* (Long Beach,

Calif.), February 23, 1955; William Gerber, "Modern Life includes toll exacted in terms of stress," *Iowa City Press-Citizen*, July 22, 1970; "Stress: modern man's silent enemy," *The Sun* (Lowell, Mass.), July 21, 1977; Patricia McCormack, "Premature aging result of modern life stress," *Chronicle Telegram* (Elyria, Ohio), March 7, 1971.

3. Terry Looker and Olga Gregson, *Stresswise: A Practical Guide for Dealing with Stress* (New York: Hodder & Stoughton, 1989), p. 26. I am indebted for this reference to Steven D. Brown, "Fighting and Fleeing: Walter Cannon and the Making of the Stressed Body," paper presented at "Making Sense of the Body" annual conference, British Sociological Association, Edinburgh, April 1998.
4. George M. Beard, *American Nervousness, Its Causes and Consequences* (New York: Putnam, 1881), p. 96.
5. Ibid., p. 98–99.
6. Silas Weir Mitchell, *Blood and Fat, and How to Make Them* (Philadelphia: J. B. Lippincott, 1877).
7. Quoted in Barbara Will, "The Nervous Origins of the American Western," *American Literature* 70, no. 2 (1998): 293–316, cited p. 301. See also Thomas Lutz, *American Nervousness, 1903* (Ithaca, N.Y.: Cornell University Press, 1991).
8. Quoted in T. J. Jackson Lears, *No Place of Grace* (New York: Pantheon Books, 1981), p. 50.
9. Among others who have talked about this historical transition from exhaustion to stress are Megan Barke, Rebecca Fribush, and Peter N. Stearns in their article "Nervous Breakdown in 20th-Century American Culture," *Journal of Social History* 3, no. 3 (2000): 565–84.
10. Walter B. Cannon, *The Mechanical Factors of Digestion* (New York: Longmans, Green & Co., 1911).
11. Walter B. Cannon, *Bodily Changes in Pain, Hunger, Fear and Rage: An Account of Recent Researches into the Function of Emotional Excitement*, 2d ed. (New York: D. Appleton, 1929), p. 41.
12. Walter B. Cannon and Daniel de la Paz, "Emotional Stimulation of Adrenal Secretion," *American Journal of Physiology* 28 (1911): 64–70.
13. Cannon, *Bodily Changes*, p. 219.
14. Walter B. Cannon and Arturo Rosenblueth, *Autonomic Neuro-effector Systems* (New York: Macmillan, 1937).
15. Walter B. Cannon, "The Role of Emotions in Disease," *Annals of Internal Medicine* 9, no. 11 (May 1936): 1453–65, cited p. 1458.
16. Ibid., p. 1455.
17. Hans Selye, *The Stress of My Life: A Scientist's Memoirs*, 2d ed. (New York: Van Nostrand Reinhold, 1979), p. 59.
18. Ibid., p. 60.
19. Hans Selye, "A Syndrome Produced by Diverse Nocuous Agents," *Nature* 138 (1936): 32.
20. Selye, *Stress of My Life*, p. 61.
21. The word "strain" is used to describe the effects of stress on the metals themselves. Properly, therefore, Selye should have used the word "strain" to describe the syndrome, and we should all today be talking about being "strained out."

   Since the word "stress" entered the English language with the new meaning given it by Selye, others have gone back and reconstructed its extended history. As early as the fourteenth century, the word "stress" was used to mean hardship or adversity. In the seventeenth century, it was brought into the sciences and given precise technical meaning through the work of the English physicist-biologist Robert Hooke ("Hooke's Law"). Hooke was interested in the varying degrees to which man-made structures, especially those made of metal, could withstand heavy loads without becoming permanently deformed. He used the word "load" to refer to the external force that a structure must resist, and the word "stress" to refer to the area over which the load impinged. "Strain" was the effect on the structure caused by the interaction of load with stress. The maximum amount of stress a material could withstand before becoming permanently deformed was referred to as its elastic limit. For more on all this, see Lawrence E. Hinkle, Jr., "The Concept of 'Stress' in the Biological and Social Sciences," *Science, Medicine, and Man* 1 (1973): 31–48.

22. Selye, *Stress of My Life*, p. xi.
23. On Cannon's concerns in particular, see Sandor Szabo, "Hans Selye and the Development of the Stress Concept: Special Reference to Gastroduodenal Ulcerogenesis," *Annals of the New York Academy of Sciences* 851 (1998):19–27. Selye also refers to Cannon's resistance to his ideas in *Stress of My Life*, in a chapter titled "When Scientists Disagree," pp. 191–92.
24. Russell Viner, "Putting Stress in Life: Hans Selye and the Making of Stress Theory," *Social Studies of Science* 29, no. 3 (1999): 391–410.
25. For more on this, see Viner, "Putting Stress in Life." Also see Hans Selye, "Implications of the Stress Concept," *New York State Journal of Medicine* 75, no. 12 (October 1975): 2139–44.
26. R. L. Engel, cited in John W. Mason, "A Historical View of the Stress Field, Part I," *Journal of Human Stress* 1 (March 1975): 10.
27. L. L. Coleman, M.D., "Your Health: Stress Tolerance Differs," *Valley Independent* (Pittsburgh, Penn.), May 21, 1973, p. 19; Alfred Toffler, *Future Shock* (New York: Random House, 1970); "The Hazards of Change," *Time*, March 1, 1971, p. 54.
28. Institute of Medicine, *Research on Stress and Human Health* (Washington, D.C.: National Academy Press, 1981), pp. 2–3.
29. For more on the way in which stress and the new expert research community concerned with diagnosing stress played out in this dispute, see S. Tesh, "The Politics of Stress: The Case of Air-Traffic Control," *International Journal of Health Services*, 14, no. 4 (1984): 569–87.
30. Roy R. Grinker and John P. Spiegel, *Men under Stress* (Philadelphia: Blackison, 1945).
31. Hudson Hoagland, "Adventures in Biological Engineering," *Science* 100 (1944): 63–64, cited in Robert Kugelman, *Stress* (Westport, Conn.: Praeger Publishers, 1992), p. 61.
32. Ibid., p. 213.
33. Viner, "Putting Stress in Life."
34. Mitchell M. Berkun, Hilton M. Bialek, Richard P. Kern, and Kan Yagi, "Experimental Studies of Psychological Stress in Man," *Psychological Monographs: General and Applied* 76, no. 15 (1962).
35. Richard S. Lazarus, "From Psychological Stress to the Emotions: A History

of Changing Outlooks," *Annual Review of Psychology* 44 (1993): 1–21, cited p. 2.

36. Thomas H. Holmes and Richard H. Rahe, "The Social Readjustment Rating Scale," *Journal of Psychosomatic Research* 11 (1967): 213–18.
37. Thomas H. Holmes and M. Masuda, "Life Change and Illness Susceptibility," in *Stressful Life Events*, ed. Barbara Snell Dohrenwend and Bruce P. Dohrenwend (New York: John Wiley & Sons, 1974), pp. 45–72, cited p. 59.
38. Robert Sward, "Personal Stress Assessment (Found Poem)," in *Four Incarnations: New and Selected Poems 1957–1991* (Minneapolis, Minn.: Coffee House Press, 1991).
39. Joseph V. Brady, "Ulcers in 'Executive' Monkeys," *Scientific American* 199 (1958): 95–100.
40. Jay M. Weiss, "Effects of Coping Behavior in Different Warning Signal Conditions on Stress Pathology in Rats," *Journal of Comparative and Physiological Psychology* 77 (1971): 1–13; J. M. Weiss, "Effects of Coping Behavior with and without a Feedback Signal on Stress Pathology in Rats," *Journal of Comparative and Physiological Psychology* 77 (1971): 22–30.
41. See Haynes Johnson, *The Age of Anxiety: McCarthyism to Terrorism* (Orlando, Fla.: Harcourt, 2006).
42. For an historical overview, see, among others, C. Kay Larson, *'Til I Come Marching Home: A Brief History of American Women in World War II* (Pasadena, Md.: Minerva Center, 1995).
43. For more on all this, see Jonathan M. Metzl, "Selling Sanity through Gender: The Psychodynamics of Psychotropic Advertising," *Journal of Medical Humanities* 24, no. 2 (2003): 79–103.
44. C. Wright Mills, *White Collar: The American Middle Classes* (New York: Oxford University Press, 1951).
45. Sloan Wilson, *The Man in the Gray Flannel Suit* (New York: Simon & Schuster, 1955).
46. William Whyte, *The Organization Man* (New York: Simon & Schuster, 1956), p. 1.
47. *The Relaxed Wife* (which has a bit of a cult status among aficionados of 1950s popular culture) has been archived on the internet and can be downloaded and viewed at no cost at Moving Image Archive, www.archive.org/movies.
48. By the 1970s, this new approach to stress had named itself the "cognitive" or "transactional" school of stress research, and was drawing freely on concepts from cybernetics as well as neobehavioral and cognitive approaches to behavior. For a summary, see S. R. Burchfield, "The Stress Response: A New Perspective," *Psychosomatic Medicine* 41 (1979): 661–72.
49. Nancy Duin and Jenny Sutcliffe, *A History of Medicine* (New York: Barnes and Noble, 1992), p. 216.
50. For more on Framingham, see "You Changed America's Heart: 1948–1998: A 50th Anniversary Tribute to the Participants in the Framingham Heart Study," at www.nhlbi.nih.gov/about/framingham/fhsbro.htm. For more on Seven Countries, see H. Toshim, Y. Koga, and H. Blackburn, eds., *Lessons for Science from the Seven Countries Study* (New York: Springer Verlag, 1994).
51. Meyer Friedman and D. Ulmer, *Treating Type "A" Behaviour and Your Heart* (London: Guild, 1985), p. 7.

52. Cited in Harris Dienstfrey, *Where the Mind Meets the Body* (New York: HarperCollins, 1991), p. 5.
53. Meyer Friedman, Ray H. Rosenman, Vernice Carroll, and Russell J. Tat, "Changes in the Serum Cholesterol and Blood Clotting Time in Men Subjected to Cyclic Variation of Occupational Stress," *Circulation* 17 (1958): 852–61.
54. Meyer Friedman and Ray H. Rosenman, "Association of Specific Overt Behavior Pattern with Blood and Cardiovascular Findings," *Journal of the American Medical Association* 169 (1959): 1286–96, cited p. 1287. Originally, Friedman and Rosenman had cast their study as an investigation of the effects of stress and emotions on the heart, but psychologists and psychiatrists, reviewing the initial grant application, had rejected it—what, after all, did cardiologists know about stress? One sympathetic colleague who had watched this happen told the researchers that if they simply removed all references in the grant application to psychological processes like stress, and instead spoke simply of a "behavior pattern" associated with their different groups (A, B, C), they were likely to be funded. The advice worked, they received the grant, the study was carried out—and the English language acquired a new term: "Type A behavior" (TAB). Gerald Friedland, "In Memoriam: Meyer Friedman, MD," *Circulation* 104 (2001): 2758.
55. R. H. Rosenman, R. I. Brand, C. D. Jenkins, M. Friedman, R. Straus, and M. Wurm, "Coronary Heart Disease in the Western Collaborative Group Study, Final Follow-up Experience of 8½ Years," *Journal of the American Medical Association* 233 (1975): 812–17.
56. T. Cooper, T. Detre, and S. M. Weiss [The Review Panel on Coronary-Prone Behavior and Heart Disease], "Coronary-prone Behavior and Coronary Heart Disease: A Critical Review," *Circulation* 63 (1981): 1199–1215.
57. Ray H. Rosenman and Margaret Chesney, "Type A Behavior Pattern: Its Relationship to Coronary Heart Disease and Its Modification by Behavioral and Pharmacological Approaches," in *Stress in Health and Disease*, ed. Michael R. Zales (New York: Brunner/Mazel, 1985), p. 209.
58. Ray H. Rosenman, "Role of Type A Behavior Pattern in the Pathogenesis of Ischemic Heart Disease, and Modification for Prevention," *Advances in Cardiology* 25 (1978): 35–46, cited p. 40. By this time, Rosenman and Friedman were no longer collaborating; but some of the best evidence at the time that modest behavioral modification could make a difference to heart healthiness had come from a study that Rosenman's former colleague had launched independently in 1977: the Recurrent Coronary Prevention Program. Focusing on patients with Type A traits who had already had at least one heart attack, the results, after four years, seemed clear: the group of patients who had received special counseling in lifestyle change had a 45 percent lower recurrence rate than two control groups (one receiving standard counseling, the second simply standard medical treatment). See Meyer Friedman, C. E. Thoresen J. J. Gill, et al., "Alteration of Type A Behavior and Its Effect on Cardiac Recurrences in Post-Myocardial Infarction Patients: Summary Results of the Recurrent Coronary Prevention Project," *American Heart Journal* 112 (1986): 653–62.
59. Rosenman, "Role of Type A Behavior Pattern," p. 41.
60. Neil Solomon, "Heart-Attack Personality: Will Success Kill 'Type A' Man?" *Los*

Angeles Times, November 16, 1979; Anonymous, "You may make a killing. Personality a heart attack sign," *Daily Times-News* (Burlington, N.C.), October 13, 1966; Rose Dostile, "Stress No. 1 Coronary Factor? Type B Better Off in a Type A World," *Los Angeles Times,* February 12, 1976); Jane E. Brody, "Rushing Your Life Away with 'Type A' Behavior," *New York Times*, October 22, 1980.

61. Elianne Riska, "The Rise and Fall of Type A Man," *Social Science and Medicine* 51 (2000): 1665–74.
62. Meyer Friedman and Ray H. Rosenman, *Type A Behavior and Your Heart* (New York: Knopf, 1974), back cover material.
63. For an example, see Judith M. Richter and Rebecca Sloan, "A Relaxation Technique," *American Journal of Nursing* 79, no. 11 (November 1979): 1960–64.
64. For some of the early seminal publications, see N. E. Miller, and L. DiCara, "Instrumental Learning of Heart Rate Changes in Curarized Rats: Shaping, and Specificity to Discriminative Stimulus," *Journal of Comparative and Physiological Psychology* 63, no. 1 (1967): 12–19; N. E Miller and A. Banuazizi, "Instrumental Learning by Curarized Rats of a Specific Visceral Response, Intestinal, or Cardiac," *Journal of Comparative and Physiological Psychology* 65 (1968): 1–7; N. E. Miller and B. R. Dworkin, "Visceral Learning: Recent Difficulties with Curarized Rats and Significant Programs for Human Research," in *Contemporary Trends in Cardiovascular Psychophysiology*, ed. P. A. Oberist et al. (Chicago: Aldine-Atherton, 1973).
65. Mary Knoblauch, "Relax! Get a Grip on your Headaches," *Chicago Tribune*, October 8, 1975.
66. D. Ragland and R. Brand, "Type A Behavior and Mortality from Coronary Heart Disease," *New England Journal of Medicine* 318, no. 2 (1988): 65–69.
67. R. B. Shekelle, R. Gale, A. M. Ostfield, et al., "Hostility, Risk of Coronary Disease, and Mortality," *Psychosomatic Medicine* 45 (1983): 219–28; T. M. Dembroski, J. M. MacDougal, R. B. Williams, et al., "Components of Type A, Hostility and Anger in Relationship to Angiographic Findings," *Psychosomatic Medicine* 47 (1985): 219–24.
68. Redford Williams, *The Trusting Heart* (New York: Times Books, 1989); Redford Williams and Virginia Williams, *Anger Kills: Seventeen Strategies for Controlling the Hostility That Can Harm Your Health* (New York: Times Books, 1993).
69. Robert Wright, "David Letterman's Cynical Heart," *Slate*, February 8, 2000, available online at www.slate.com/id/74493/.
70. Robert Ader, "On the Development of Psychoneuroimmunology," *European Journal of Pharmacology* 405 (September 29, 2000): 167–76, cited p. 168. The reference for the landmark article itself is Robert Ader and Nicholas Cohen, "Behaviorally Conditioned Immunosuppression," *Psychosomatic Medicine* 37, no. 4 (July–August 1975): 333–40. Ader later discovered—and frequently called attention to the fact—that classical conditioning of the immune system had been successfully carried out by several Russian investigators as early as the 1920s but had received little, if any, attention in the English-speaking scientific community.
71. J. E. Blalock and E. M. Smith, "Human Leukocyte Interferon: Structural and Biological Relatedness to Adrenocorticotropic Hormone and Endorphins," *Proceedings of the National Academy of Sciences* 77 (1980): 4597–5972; J. M. Williams, R. G. Peterson, P. A. Shea, J. F. Schmedtje, D. C. Bauer, and D. L. Felten,

"Sympathetic Innervation of Murine Thymus and Spleen: Evidence for a Functional Link between the Nervous and Immune Systems," *Brain Research Bulletin* 6 (1981): 83–94; for a later review, see W. L. Farrar, "Evidence for the Common Expression of Neuroendocrine Hormones and Cytokines in the Immune and Central Nervous Systems," *Brain, Behavior, Immunity* 2 (1988): 322–27.

72. Robert Ader, "Presidential Address: Psychosomatic and Psychoimmunologic Research," *Psychosomatic Medicine* 42 (1980): 307–22. A year later, in 1981, Ader brought out the first overview of the then-available evidence for neuroimmune communication (including his own work) in an edited volume he boldly titled *Psychoneuroimmunology* (New York: Academic Press, 1981).
73. Emily Martin, *Flexible Bodies: Tracking Immunity in American Culture from the Days of Polio to the Age of AIDS* (Boston: Beacon Press, 1994), p. 49.
74. Randy Shilts, *And the Band Played On* (New York: St. Martin's Press, 1987), p. 347.
75. George Solomon and Lydia Temoshok, "A Psychoneuroimmunologic Perspective on AIDS," in *Psychosocial Perspectives on AIDS*, ed. Lydia Temoshok and Andrew Baum (Hillsdale, N.J.: Lawrence Erlbaum, 1990), p. 234.
76. Jason Serinus, ed., *Psychoimmunity and the Healing Process* (Berkeley, Calif.: Celestial Arts, 1986), p. 83.
77. A lot of the pioneering work here was carried out by an American husband-and-wife team at Ohio State University, Janice Kiecolt-Glaser (a psychologist) and Ronald Glaser (an immunologist). See, for example, Ronald Glaser, Janice K. Kiecolt-Glaser, R. H. Bonneau, et al., "Stress-Induced Modulation of the Immune Response to Recombinant Hepatitis B Vaccine," *Psychosomatic Medicine* 54, no. 1 (1992): 22–29; Janice K. Kiecolt-Glaser, Ronald Glaser, S. Gravenstein, et al., "Chronic Stress Alters the Immune Response to Influenza Virus Vaccine in Older Adults," *Proceedings of the National Academy of Sciences* 93 (1996): 3043–47; Janice K. Kiecolt-Glaser, Ronald Glaser, J. T. Cacioppo, et al., "Marital Conflict in Older Adults: Endocrinological and Immunological Correlates," *Psychosomatic Medicine* 59 (1997): 339–49; Ronald Glaser, Janice K. Kiecolt-Glaser, W. B. Malarkey, and J. F. Sheridan, "The Influence of Psychological Stress on the Immune Response to Vaccines," *Annals of the New York Academy of Sciences* 840 (1998): 649–55. For a more general review of the new work on stress and immune functioning from the mid-1990s, see A. J. Dunn, "Psychoneuroimmunology: Introduction and General Perspectives" in *Stress, the Immune System, and Psychiatry*, ed. B. Leonard and K. Miller (New York: John Wiley & Sons, 1995).
78. Interview by Anne Harrington with Spiegel Breast Cancer support group, Stanford Medical School, September 13, 1995. All women who participated in these interviews signed consent forms that had been approved by the Stanford Medical School human subjects review board.

Chapter Five: Healing Ties

1. "Antidote to Loneliness," sermon given by Rev. Dr. Gordon Moyes to the Wesley Mission, Sydney, Australia, on September 23, 2001, archived on the mission website at www.wesleymission.org.au/ministry/sermons/010923.html.

2. Geoffrey Cowley with Anne Underwood, "Is Love the Best Drug?" *Newsweek*, March 16, 1998.
3. Émile Durkheim, *Le Suicide: Étude de sociologie* (Paris: F. Alcan, 1897); translated as Émile Durkheim, *Suicide: A Study in Sociology* (Glencoe, Ill: Free Press, 1951).
4. P. Besnard, "Anomie," in *International Encyclopedia of the Social and Behavioral Sciences*, editors in chief Neil J. Smelser and Paul B. Baltes (New York: Elsevier, 2001), pp. 510–13.
5. C. Stout, J. Marrow, E. N. Brandt, Jr., and S. Wolf, "Unusually Low Incidence of Death from Myocardial Infarction: Study of an Italian-American community in Pennsylvania," *Journal of the American Medical Association* 188 (1964): 845–49. The Stout study was received skeptically by at least some epidemiologists, who suggested that the population size studied was too low to allow meaningful statistical conclusions and pointed to various confounds in the data. Nevertheless, it catalyzed a great deal of further investigation into the Roseto phenomenon.
6. Stewart Wolf and Harold Wolff, *Human Gastric Function: An Experimental Study of a Man and His Stomach*, with a foreword by Walter B. Cannon (New York: Oxford University Press, 1943).
7. J. T. Doyle, T. R. Dawber, W. B. Kannel, A. S. Heslin, and H. A. Kahn, "Cigarette Smoking and Coronary Heart Disease: Combined Experience of the Albany and Framingham Studies," *New England Journal of Medicine* 266 (1962): 796–801; Ancel Keys, C. Aravanis, H. W. Blackburn, F. S. P. Van Buchem, et al., "Epidemiologic Studies Related to Coronary Heart Disease: Characteristics of Men Aged 40–59 in Seven Countries," *Acta Medica Scandinavica Supplementum* 460 (1967): 1–392.
8. Joel Greenberg, "The Americanization of Roseto," *Science News* 113, no. 23 (June 10, 1978): 378–80, cited p. 380.
9. Stewart Wolf and John G. Bruhn, *The Power of Clan* (New Brunswick, N.J.: Transaction Publishers, 1993), p. 10.
10. John G. Bruhn and Stewart Wolf, *The Roseto Story* (Norman: University of Oklahoma Press, 1979), pp. 80, 81–82.
11. Wolf and Bruhn, *The Power of Clan*.
12. B. Egolf, J. Lasker, S. Wolf, and L. Potvin, "The Roseto Effect: A 50–Year Comparison of Mortality Rates," *American Journal of Public Health* 82, no. 8 (August 1992): 1089–92, cited p. 1089. The quote here from a young Rosetan is taken from the essay by Greenberg, "Americanization of Roseto," p. 379. For a sampling of more interview material, see J. G. Bruhn, B. Phillips, and S. Wolf, "Social Readjustment and Illness Patterns: Comparisons between First, Second and Third Generation Italian-Americans Living in the Same Community," *Journal of Psychosomatic Research* 16 (1972): 387–94.
13. Greenberg, "Americanization of Roseto," p. 379.
14. Ibid.
15. Stewart Wolf, "Visceral Responses to the Social Environment," *Acta Physiologica Scandinavica Supplementum* 640 (1997):140–43.
16. S. Leonard Syme, "Historical Perspective: The Social Determinants of Disease—Some Roots of the Movement," *Epidemiologic Perspectives & Innovations* 2, no. 2 (2005), available online at www.epi-perspectives.com/content/2/1/2.
17. Michael Gideon Marmot, "Acculturation and Coronary Heart Disease in

Japanese-Americans," Ph.D. diss., University of California, Berkeley, 1975; published in part as Michael Gideon Marmot and S. Leonard Syme, "Acculturation and CHD in Japanese-Americans," *American Journal of Epidemiology* 104 (1976): 225–47.

18. Syme, "Historical Perspective."
19. J. A. Barnes, "Class and Committees in a Norwegian Island Parish," *Human Relations* 7 (1954): 39–58. Other influential studies in this tradition include J. V. Mitchell's edited collection, *Social Networks in Urban Situations: Analyses of Personal Relationships in Central African Towns* (Manchester: Manchester University Press, 1969), and Mark Granovetter's 1973 paper, "The Strength of Weak Ties," *American Journal of Sociology* 78, no. 6 (1973): 1360–80.
20. Interview by Anne Harrington with Lisa Berkman, June 9, 2000.
21. Lisa F. Berkman and S. Leonard Syme, "Social Networks, Host Resistance, and Mortality: A Nine-Year Follow-up Study of Alameda County Residents," *American Journal of Epidemiology* 109, no. 2 (1979): 186–204. For some further background to the Alameda County study and Berkman's use of its data, see Robin M. Henig, *People's Health* (Washington, D.C.: National Academy Press, 1996).
22. For the classic overview of all these studies, see James S. House, K. R. Landis, et al., "Social Relationships and Health," *Science* 241, no. 4865 (1988): 540–45.
23. Sidney Cobb, "Social Support as a Moderator of Life Stress," *Psychosomatic Medicine* 38 (1976): 300–314.
24. John Cassel, "The contribution of the Social Environment to Host Resistance," *American Journal of Epidemiology* 104 (1976): 107–23, cited p. 113. See also John Cassel, "Psychosocial Processes and Stress: Theoretical Formulations," *International Journal of Health Services* 4 (1974): 471–82.
25. The increase in articles with the words "social support" in their titles was reported by James S. House, D. Umberson, and K. R. Landis, "Structures and Processes of Social Support," *American Review of Sociology* 14 (1988): 293–318, cited pp. 293–94.
26. Alfred Fusco, "Friends Can Be Good Medicine," *Syracuse Post-Standard*, August 6, 1984.
27. Carol Turkington, "Have You Hugged Your Immune System Today?" cover of *Self*, October 1998.
28. Robert D. Putnam, *Bowling Alone: The Collapse and Revival of American Community* (New York: Simon & Schuster, 2000).
29. Michael Gideon Marmot, "Social Differentials in Health within and between Populations," in *Health and Wealth*, published as a special volume of *Daedalus: Proceedings of the American Academy of Arts and Sciences* 123, no. 4 (1994): 197–216, cited p. 200.
30. Michael Gideon Marmot, G. D. Smith, S. Stansfeld, C. Patel, et al., "Health Inequalities among British Civil Servants: The Whitehall II Study," *Lancet* 337 (1991): 1393–97.
31. Ichiro Kawachi and Bruce P. Kennedy, *The Health of Nations: Why Inequality Is Harmful to Your Health* (New York: New Press, 2002), p. 37.
32. Cited in *Social Determinants of Health*, ed. Michael Marmot and Richard G. Wilkinson (New York: Oxford University Press, 2006), p. 166.
33. James J. Lynch, *The Broken Heart* (New York: Basic Books, 1977). pp. 11, 13.

34. Ibid., pp. 65–66.
35. Amy Schoenfeld, an honors student at Harvard College at the time of this writing, did a citation search of *The Broken Heart* and found that, of the 248 articles that have cited the book since its publication in 1977, only 8 percent were from general/internal medicine journals and 16 percent were from psychiatry journals. Put another way, fully 76 percent of all references to Lynch have been made in nonmedical—psychology, environmental health, social work, family studies, and social science—journals. See Amy Schoenfeld, "Don't Go Breaking My Heart: A Post–World War II History of Loneliness as a Risk Factor for Cardiovascular Disease," senior honors thesis, Department of the History of Science, Harvard University, 2007.
36. James J. Lynch, "Warning: Living Alone Is Dangerous to Your Health," *U.S. News & World Report,* June 30, 1980, pp. 47–48.
37. H. Jack Geiger, "Love and/or Die," review of *The Broken Heart*, *New York Times*, July 10, 1977.
38. For more on this, see the 2007 senior honors thesis of Amy Schoenfeld, "Don't Go Breaking My Heart: A Post–World War II History of Loneliness as a Risk Factor for Cardiovascular Disease."
39. Louise Bernikow, "Alone: Yearning for Companionship in America," *New York Times Magazine*, August 15, 1982. Four years later, Bernikow wrote a book on this topic, *Alone in America* (New York: Harper & Row, 1986).
40. Lynch, *Broken Heart,* p. 84.
41. Hansi Kennedy, "Memories of Anna Freud," *American Imago* 53, no. 3 (1996): 205–09.
42. John Bowlby, *Maternal Care and Mental Health* (Geneva: World Health Organization, 1951), cited in Cara Flanagan, *Early Socialisation* (New York: Routledge, 1999), pp. 3–4.
43. René A. Spitz, *The First Year of Life: A Psychoanalytic Study of Normal and Deviant Development of Object Relations*, in collaboration with W. Godfrey Cobliner (New York: International Universities Press, 1965), p. 278.
44. René Spitz, "Hospitalism: An Inquiry into the Genesis of Psychiatric Conditions in Early Childhood," parts 1 and 2, *Psychoanalytic Studies of the Child* 1 (1945): 53–74, and 2 (1945): 113–17. The idea of a link between physical growth and emotional deprivation and/or emotional abuse was later called deprivation dwarfism, or psychosocial dwarfism. For a classic overview, written from a psychoanalytic perspective, see Lytt Gardner, "Deprivation Dwarfism." *Scientific American* 227, no. 1 (1972): 76–82.
45. E. M. Widdowson, "Mental Contentment and Physical Growth," *Lancet* 1 (1951): 1316–18.
46. David Tenenbaum, quoting Mary Carlson, in "The Science of Mother's Day," available online at http://whyfiles.org/087mother/4.html.
47. Douglas Martin, "Elsie Widdowson, 93, a Pioneer in Nutrition," *New York Times*, June 26, 2000. Two years later, in 2002, health psychologist Shelley Taylor opened her book *The Tending Instinct* with a detailed account of Widdowson's two orphanages, and made a similar point: "The important story was not about food." Shelley E. Taylor, *The Tending Instinct* (New York: Owl Books, 2002), p. 8.
48. T. George Harris, "Heart and Soul: Out on the Cutting Edge, Hard-nosed

Researchers Study the Tie between Healthy Emotions and Stout Hearts," *Psychology Today*, January–February 1989, available online at http://findarticles.com/p/articles/mi_m1175/is_n1_v23/ai.7049288.

49. Dean Ornish, *Love & Survival*. The quote here is from a paperback edition with the subtitle "*8 Pathways to Intimacy and Health*" (New York: HarperCollins, 1999), pp. 2–3. Ornish published an initial version of his argument as early as 1990; see Dean Ornish, S. E. Brown. L. W. Scherwitz, et al., "The Healing Power of Love," *Prevention*, February 1991, pp. 60–66, 135–141.
50. "Study Finds No Reduction in Deaths or Heart Attacks in Heart Disease Patients Treated for Depression and Low Social Support," news release, National Heart, Lung, and Blood Institute, November 12, 2001, available online at www.nhlbi.nih.gov/new/press/01-11-13.htm.
51. Lisa Berkman, guest presentation on November 20, 2006, History of Science 177, "Stories under the Skin: The Mind-Body Connection in Modern Medicine" (instructor, Anne Harrington), Harvard Faculty of Arts and Sciences.
52. Bernie Siegel, *Love, Medicine, and Miracles* (1986; New York: HarperPerennial, 1990), p. 180.
53. D. Oken, "What to Tell Cancer Patients: A Study of Medical Attitudes," *Journal of the American Medical Association* 175 (1961): 1120–28.; D. H. Novack, R. Plumer, R. L. Smith, H. Ochitill, et al., "Changes in Physicians' Attitudes Toward Telling the Cancer Patients," *Journal of the American Medical Association* 241 (1979): 897–900. Both these sources were cited in Jimmie C. Holland, "History of Psycho-oncology: overcoming attitudinal and conceptual barriers," *Psychosomatic Medicine* 64 (2002): 206–221.
54. Cited in Holland, "History of Psycho-oncology," p. 216.
55. Cited in Lynn Smith, "Cancer Patients Mind over Body: Doubt Rekindled," *Los Angeles Times*, August 20, 1985.
56. David Spiegel, personal communication.
57. Irvin Yalom, *Existential Psychotherapy* (New York: Basic Books, 1980), p. 45.
58. David Spiegel, *Living Beyond Limits* (New York: Random House, 1994), chapter 6, "Detoxifying Dying," excerpt available online at http://med.stanford.edu/school/Psychiatry/PSTreatlab/lbltext.html.
59. David Spiegel, J. R. Bloom, and Irwin D. Yalom, "Group Support for Patients with Metastatic Cancer: A Randomized Prospective Outcome Study," *Archives of General Psychiatry* 38 (1981): 527–33.
60. Scott Winokur, "Psychotherapy Helps Cancer Patients," *San Francisco Examiner*, reprinted in *The News* (Frederick, Md.), December 11, 1989.
61. David Spiegel, J. R. Bloom, H. C. Kraemer, and E. Gottheil, "Effect of Psychosocial Treatment on Survival of Patients with Metastatic Breast Cancer," *Lancet* 2, no. 8668 (1989): 888–91.
62. Anonymous, "Psychosocial Intervention and the Natural History of Cancer," *Lancet* 334, no. 8668 (1989): 901.
63. Heather Goodare, "A Scientific Pioneer of Cancer Groups," *Advances: The Journal of Mind-Body Health* 10, no. 4 (1994): 71–73.
64. Janny Scott, "Study Says Cancer Survival Rises with Group Therapy," *Los Angeles Times*, May 11, 1989.
65. "Therapeutic Support Groups," interview by Bill Moyers with David Spiegel,

Healing and the Mind 1993 by Public Affairs Television, Inc. and David Grubin Productions, Inc. Permission to print granted from Bill Moyers and Doubleday. Available online at http://med.stanford.edu/school/Psychiatry/PSTreatLab/moyer.html.

66. Lauren John, "How Support Groups Affect Survival," *Breast Cancer Action Newsletter* 56 (October–November 1999), available online at www.bcaction.org/Pages/SearchablePages/1999Newsletters/Newsletter056E.html.
67. Goodare, "Scientific Pioneer."
68. David Spiegel, "Social Support: How Friends, Family, and Groups Can Help," in *Mind Body Medicine*, ed. Daniel Goleman and Joel Gurin (Yonkers, N.Y.: Consumer Report Books, 1993).
69. Pamela J. Goodwin, Molyn Leszcz, Marguerite Ennis, Jan Koopmans, et al., "The Effect of Group Psychosocial Support on Survival in Metastatic Breast Cancer," *New England Journal of Medicine* 345, no. 24 (2001): 1719–26.
70. *USA Today*, December 12, 2001, available online at www.usatoday.com/news/health/cancer/2001-12-12-breast-cancer-therapy.htm.
71. Cited in Larry Lachman, "Group Therapy for All Cancer Patients," *Psychology Today* 35, no. 5 (2002): 27.
72. On July 23, 2007, as this book was going to press, an "early view" version of an article by Spiegel and his colleagues, due to be published in the medical journal *Cancer,* was posted online. In the article, Spiegel and his colleagues did finally concede that, after fourteen years, they had failed to replicate the "earlier finding that intensive group therapy extended survival time of women with metastatic breast cancer." They offered various suggestions as to why the replication might have failed, including the fact that all cancer patients today have many more opportunities to seek and experience psychosocial support than had been the case several decades ago. See David Spiegel, Lisa D. Butler, Janine Giese-Davis, Cheryl Koopman, Elaine Miller, Sue DiMiceli, Catherine Classen, Patricia Fobair, Robert W. Carlson, and Helena Kraemer, "Effects of Supportive-Expressive Group Therapy on Survival of Patients with Metastatic Breast Cancer: A Randomized Prospective Trial," *Cancer* (July 23, 2007), available online at http://dx.doi.org/10.1002/cncr.22890, copyright © 2007 American Cancer Society.
73. Cited in Tori DeAngelis, "How Do Mind-Body Interventions Affect Breast Cancer?" *Monitor on Psychology* 33, no. 6 (June 2002), available online at www.apa.org/monitor/jun02/mindbody.html.
74. Spiegel Breast Cancer support group, Stanford Medical School, September 13, 1995, group interview with Anne Harrington. All women who participated in these interviews signed consent forms, as required by the Stanford Medical School human subjects review board.

Chapter Six: Eastward Journeys

1. Daniel Goleman, "Finding Happiness: Cajole Your Brain to Lean to the Left," *New York Times*, February 4, 2003.
2. Edward Said, *Orientalism* (New York: Pantheon Books, 1978).
3. Arthur Christy, *The Orient in American Transcendentalism: A Study of Emerson,*

Thoreau, and Alcott (New York: Columbia University Press, 1931), pp. 12–13 and 38–46.

4. For some perspectives on the history of Theosophy, see Joy Dixon, *Divine Feminine: Theosophy and Feminism in England*, The Johns Hopkins University Studies in Historical and Political Science (Baltimore, Md.: Johns Hopkins University Press, 2001); Sylvia Cranston, *HPB: The Extraordinary Life and Influence of Helena Blavatsky, Founder of the Modern Theosophical Movement* (New York: Putnam, 1993); Stephen Prothero, *The White Buddhist: The Asian Odyssey of Henry Steel Olcott* (Bloomington: Indiana University Press, 1996).
5. One of the best general analyses of the transformations of Orientalism in post-colonial times is Richard King, *Orientalism and Religion: Post-Colonial Theory, India and "the Mystic East"* (New York: Routledge, 1999).
6. Cover of *Saturday Evening Post*, May 4, 1968.
7. "The Beatles Anthology," transcripted material © 1995 by American Broadcasting Companies, accessed October 2, 2004, at http://homepages.tesco.net/~d.saunders/TMCafe/Beatles.html. (This link is no longer live.)
8. Anonymous, "Hinduism in New York: A Growing Religion," *New York Times*, November 2, 1967.
9. Alan Watts, *Behold the Spirit* (1947; New York: Pantheon Books, 1971), p. xxi.
10. Carole Tonkinson, ed., *Big Sky Mind: Buddhism and the Beat Generation* (New York: Riverhead Books, 1995).
11. Barney Lefferts, "Chief Guru of the Western World," *New York Times*, December 17, 1967.
12. The title "Maharishi" was bestowed on him during his early career as a teacher in India. His biographer, Paul Mason, notes that it was used for the first time in a 1955 publication of teachings titled *Beacon Light of the Himalayas*. Paul Mason, *The Maharishi* (Rockport, Mass.: Element Books, 1994).
13. M. Ebon, *Maharishi* (London: Pearl, 1967), p. 3.
14. Lefferts, "Chief Guru."
15. Maharishi Mahesh Yogi, *Meditations of Maharishi Mahesh Yogi* (New York: Bantam Books, 1973), pp. 17–18.
16. See *Encyclopedia of the American Religious Experience: Studies of Traditions and Movements*, ed. Charles Lippy and Peter Williams (New York: Scribner's, 1988), vol. 2, pp. 693–95.
17. Reported by William Sims Bainbridge, *Sociology of Religious Movements* (New York: Routledge, 1997), pp. 187–91, as cited in "Religious Movements: Transcendental Meditation," available online at http://religiousmovements.lib.virginia.edu/nrms/tm.html.
18. Lefferts, "Chief Guru."
19. Mason, *Maharishi*, p. 137.
20. See John Lennon and Jann S. Wenner, *Lennon Remembers* (New York: Da Capo Press, 2000), p. 27.
21. This strain of thinking in the 1970s peaked with the popular 1975 work of Fritjof Capra, *The Tao of Physics* (Berkeley, Calif.: Shambala Publications). For some critical perspectives, see Sal P. Restivo, "Parallels and Paradoxes in Modern Physics and Eastern Mysticism: I—A Critical Reconnaissance," *Social Studies of*

Science 8, no. 2 (1978): 143–81, and "Parallels and Paradoxes in Modern Physics and Eastern Mysticism: II—A Sociological Perspective on Parallelism," *Social Studies of Science* 12, no. 1 (1982): 37–71.

22. Lawrence J. Domash, "The Transcendental Meditation Technique and Quantum Physics," in *Scientific Research on Maharishi's Transcendental Meditation and TM-Sidhi Program*, vol. 1, ed. David W. Orme-Johnson and John T. Farrow (West Germany: Maharishi European University Press, 1977), pp. 652–70.
23. Robert Keith Wallace, "Physiological Effects of Transcendental Meditation" *Science* 167 (1970): 1751–54, cited p. 1754. See also Robert Keith Wallace, "Proposed Fourth Major State of Consciousness," in *Scientific Research on Maharishi's Transcendental Meditation and TM-Sidhi Program*, vol. 1, ed. David W. Orme-Johnson and John T. Farrow (West Germany: Maharishi European University Press, 1977), pp. 43–78.
24. Maggie Scarf, "Tuning Down with TM," *New York Times*, February 9, 1975.
25. Herbert Benson, "Mind-Body Pioneer," *Psychology Today*, May–June, 2001, available online at www.psychologytoday.com/htdocs/prod/PTOArticle/PTO-20010501-000026.ASP (accessed September 4, 2004).
26. Herbert Benson, David Shapiro, Bernard Tursky, and Gary E. Schwartz, "Decreased Systolic Blood Pressure through Operant Conditioning Techniques in Patients with Essential Hypertension," *Science* 173, no. 998 (August 20, 1971): 740–42.
27. Herbert Benson, "Yoga for Drug Abuse," *New England Journal of Medicine* 281, no. 20 (1969): 1133.
28. Benson, "Mind-Body Pioneer."
29. Eugene Taylor, "A Perfect Correlation Between Mind and Brain: William James's *Varieties* and the Contemporary Field of Mind/Body Medicine," *Journal of Speculative Philosophy* 17, no. 1 (2003): 40–52, cited p. 45.
30. Interview by Anne Harrington with Herbert Benson, Harvard University, May 18, 1998.
31. Edward B. Fiske, "Thousands Finding Meditation Eases Stress of Living," *New York Times*, December 11, 1972.
32. Wade Roush, "Herbert Benson: Mind-Body Maverick Pushes the Envelope," *Science* New Series 276, no. 5311 (April 18, 1997): 357–59.
33. Interview by Anne Harrington with Herbert Benson, Harvard University, May 18, 1998.
34. Herbert Benson, J. F. Beary, and M. P. Carol, "The Relaxation Response," *Psychiatry* 37 (1974): 37–46.
35. The term caught on. In 1993, a widely cited survey of unconventional therapy use in the United States found that "relaxation" techniques topped the list. See David M. Eisenberg, R. C. Kessler, C. Fsterm, F. E. Norlock, et al., "Unconventional Medicine in the United States: Prevalence, Costs, and Patterns of Use," *New England Journal of Medicine* 328, no. 4 (1993): 246–52.
36. Cover of *Time* magazine, October 13, 1975.
37. Herbert Benson, *The Relaxation Response* (1975; New York: Avon Books, 1976), p. 117.
38. Scarf, "Tuning Down with TM."

39. William Nolen, M.D., quoted on the back jacket of Herbert Benson, *The Relaxation Response* (New York: Morrow, 1975).
40. For more on the general history of Buddhism in America, see the various essays in *The Faces of Buddhism in America*, eds. Charles S. Prebish and Kenneth K. Tanaka (Berkeley: University of California Press, 1998), and *Westward Dharma: Buddhism beyond Asia*, ed. Charles S. Prebish and Martin Baumann (Berkeley: University of California Press, 2002).
41. Jon Kabat-Zinn, *Full Catastrophe Living: Using the Wisdom of Your Body and Mind to Face Stress, Pain, and Illness* (New York: Delacorte, 1990), p. 11.
42. Richard Streitfeld, "Mindful Medicine: An Interview with Jon Kabat-Zinn," *Primary Point* 8, no. 2 (Summer 1991), available online at www.kwanumzen.com/primarypoint/ (accessed September 12, 2004).
43. Jon Kabat-Zinn, "An Out-Patient Program in Behavioral Medicine for Chronic Pain Patients Based on the Practice of Mindfulness Meditation: Theoretical Considerations and Preliminary Results," *General Hospital Psychiatry* 4 (1982): 33–47; J. Kabat-Zinn, L. Lipworth, and R. Burney, "The Clinical Use of Mindfulness Meditation for the Self-Regulation of Chronic Pain," *Journal of Behavioral Medicine* 8, no. 2 (1985): 163–90.
44. J. D. Bernhard, J. Kristeller, and J. Kabat-Zinn, "Effectiveness of Relaxation and Visualization Techniques as a Adjunct to Phototherapy and Photochemotherapy of Psoriasis," *Journal of the American Academy of Dermatology* 19, no. 3 (1988): 572–74; J. Kabat-Zinn, E. Wheeler, T. Light, A. Skillings, et al., "Influence of a Mindfulness-Based Stress Reduction Intervention on Rates of Skin Clearing in Patients with Moderate to Severe Psoriasis Undergoing Phototherapy (UVB) and Photochemotherapy (PUVA)," *Psychosomatic Medicine* 60, no. 5 (1998): 625–32.
45. The study in question showed a difference in both right to left brain shift in the prefrontal cortex (indicating improved affect) immune response to a flu vaccine between the group that received training and a wait-list control group. It did not, however, demonstrate a relationship between how frequently people actually practiced the meditation and these brain and immune responses. See Richard J. Davidson, J. Kabat-Zinn, J. Schumacher, M. Rosenkranz, et al. "Alterations in Brain and Immune Function Produced by Mindfulness Meditation," *Psychosomatic Medicine* 65 (2003): 564–70.
46. Kabat-Zinn, *Full Catastrophe Living*.
47. Email to author, February 23, 2007.
48. For an historical and sociological analysis of the making of the television series and its impact, see Navaz Percy Kananjia, "Healing and the Mind: The Politics of Popular Media and the Development of Mind-Body Medicine in 1990s America," senior honors thesis, Department of the History of Science, Harvard University, 2001.
49. For the transcript of these conversations, see David Eisenberg, "The Mystery of Chi: Medicine in a Mind/Body Culture and Another Way of Seeing," in *Healing and the Mind*, ed. Bill Moyers (New York: Doubleday, 1993), pp. 251–314.
50. E.g., Mark Harrison, "Star Wars and the Seraglio," written in 1999 and posted on August 22, 2005, on the blog "Taiwan/China: Theory/Futures": http://mharrison.wordpress.com/2005/08/22/star-wars-and-the-seraglio-1999–2/.

51. Christopher Simmons, ed., "Qigong, 'the Asian Miracle from China,' Can Now Easily Be Learned with New Instructional Video from Qigong Expert Sal Alfarone," press release from June 3, 2003, on behalf of Sal Alfarone's video "Secrets of China," availabe online at www.send2press.com/PRnetwire/pr_03_0603–qigong.shtml.
52. Prior to the Bill Moyers series, there had been some modest interest in qigong practices in the United States, especially in northern California, where there was a large Chinese immigrant population. The first acupuncture schools had been set up there in the 1970s, and San Franciso became, in 1988, the first city to house a Qigong Institute (which was affiliated with the East West Academy of Healing Arts, in existence since 1973). The First World Congress on Qigong was held in Berkeley, California, in 1990.
53. John M. Glionna, "Turning East for the Answers to Medical Mysteries," *Los Angeles Times*, September 17, 1996.
54. Indeed, in 1929 a proposal was put forward by two Western-trained Chinese physicians that, for the sake of the people's health, practice of the old methods should be outlawed outright. The "old medicine is still conning the people with its charlatan, shamanic, and geomancing ways," they wrote. "For every day that old medicine is not abolished, the people's ideas will not change, the cause of modern medicine will not be able to progress, and sanitary measures will not be able to advance." The proposal met with great approval at the government-sponsored meeting where it was first presented. It was never formalized into law only because the traditional practitioners and pharmacies—many of whom maintained lucrative businesses both internal to China and with Chinese clients living abroad—orchestrated massive protests and strikes. Still, by the 1930s a consensus was emerging that the best way, long term, to ensure the survival of the old ways was through modernization—a process that was called "Sino-Western Convergence and Intercourse." See Bridie J. Andrews, "Tuberculosis and the Assimilation of Germ Theory in China," *Journal of the History of Medicine* 52 (1997): 114–57, especially p. 142.
55. Heiner Fruehauf, "Science, Politics, and the Making of 'TCM': Chinese Medicine in Crisis," *Journal of Chinese Medicine* 61 (October 1999): 1–9, cited p. 3.
56. Ibid., p. 2.
57. Ibid.; Peggy Durdin, "Medicine in China: A Revealing Story," *New York Times*, February 28 1960.
58. James Reston, "Now Let Me Tell You about My Appendectomy in Peking . . ." *New York Times*, July 26, 1971.
59. Ibid.
60. Ibid.
61. Herbert Burkholz, "Pain: Solving the Mystery; New Insights into the Nature of Pain Have Led to Better Ways of Controlling It," *New York Times*, September 27, 1987.
62. E. Grey Dimond, "Acupuncture Anesthesia: Western Medicine and Chinese Traditional Medicine," *Journal of the American Medical Association* 218 (1971): 1558–63.
63. Gary Kaplan, "A Brief History of Acupuncture's Journey to the West," *Journal of American Chinese Medicine* 3 (1997): S5–S10. The comment was first published,

to my best knowledge, in an article titled " 'I have seen acupuncture work,' says Nixon's doctor," *Today's Health* 50 (July 1972): 50–56, cited p. 51. I have not seen the original source.

64. For more on Eisenberg's original experiences in China, see his memoir: David Eisenberg, with Thomas Lee Wright, *Encounters with Qi: Exploring Chinese Medicine* (New York: Penguin, 1987).
65. Kunio Miura, "The Revival of Qi: Qigong in Contemporary China," in *Taoist Meditation and Longevity Techniques*, ed. Livia Kohn and Yoshinobu Sakade (Ann Arbor: University of Michigan Center for Chinese Studies, 1989), pp. 331–62.
66. For more information on this set of complex developments, see, among others, Elizabeth Hsu, *The Transmission of Chinese Medicine* (Cambridge: Cambridge University Press, 1999); Benjamin Perry, "*Qigong*, Daoism and Science: Some Contexts for the *Qigong* Boom," in *Modernization of the Chinese Past*, ed. Mabel Lee and A. D. Syrokomla-Stefanowska (Broadway, Australia: Wild Peony, 1993), pp. 166–79.
67. For a sampling of media coverage of the 1979 U.S. visit, see John Dart, "Dalai Lama to Visit L.A. in September: Plans Four Public Addresses during Five Days in Southland," *Los Angeles Times*, August 22, 1979; "Dalai Lama Arrives for Tour of U.S.," *New York Times*, September 4, 1979; "Dalai Lama to Speak at Constitution Hall," *Washington Post*, September 7, 1979; George Vecsey, "Reporter's Notebook: To Many, Dalai Lama's Visit Is Dream Come True: Worlds Meshing Gently," *New York Times*, September 10, 1979; Amanda Russell, "Los Angeles Visit of the Dalai Lama," *Los Angeles Times*, September 15, 1979; "The Visitor from Tibet," *New York Times*, September 24, 1979.
68. Jeffery Paine, "Dalai Lama Suburbia," *Boston Globe*, September 14, 2003.
69. For more on the immediate politics leading to the 1979 visit, see John Powers, *Engaged Buddhism in the West* (Somerville, Mass.: Wisdom Publications, 2000).
70. John Dart, "Tibet's Exiled Dalai Lama to Visit U.S. This Fall," *Los Angeles Times*, January 21, 1979.
71. George Vecsey, "Dalai Lama, in Texas, Hears Debate among Learned Men," *New York Times*, September 20, 1979.
72. For coverage of the Harvard visit, see Elizabeth E. Ryan, "Hello Dalai: East Meets West," *Harvard Crimson*, October 24, 1979, and Susan K. Brown, "A Lama on Wheels," *Harvard Crimson*, October 20, 1979.
73. Cited in William J. Cromie, "Meditation Changes Temperatures: Mind Controls Body in Extreme Experiments," *Harvard University Gazette*, April 18, 2002, available online at www.hno.harvard.edu/gazette/2002/04.18/09=tummo.html.
74. Jeffery Paine, *Re-enchantment* (New York: W. W. Norton, 2004), p. 202. Benson's private notes of the 1979 meeting read: "D[alai] L[ama] said our studies might convince his neighbors to the East (China) of the worth of [the Tibetan] religion." (10/18/79, 2:30–3:15 at Dana Palmer House at Harvard, with Herbert Benson, HHDL, Robert Allen, Henry Meadow, Ellen Epstein [photography], Mark Epstein, Joan Borysenko, Myrin Borysenko.) A 1982 letter from Benson to the Dalai Lama reminds the Tibetan leader of the political goal motivating his interest in allowing scientific studies of Tibetan meditation in the course of seeking his advice as to whether or not to accept an invitation to travel to China to investigate qigong practices: "Do you feel this proposed trip would be an oppor-

tunity for me to communicate my scientific findings related to g Tum-mo to the Chinese? Is this what you had in mind when we first met and you permitted me to study your religious practices? It is my recollection that you said my findings might help convince 'our friends to the East' of the practical worth of Tibetan Buddhism." (Letter to His Holiness the Dalai Lama, November 18, 1982.) The Dalai Lama responded by encouraging Benson to travel to China and tell the Chinese about some of his scientific findings relevant to Tibetan religious practices. (Response December 6, 1982, by Tempa Tseringm Deputy Secretary in the Office of His Holiness the Dalai Lama.) All these notes and letters can be found in box 2, folder 28, "Dalai Lama & Dr. Yeshi Donden's Visit, 1979–1987, n.d.," MG. 898, Herbert Benson Archives, Countway Medical School.

75. For the study that claimed to show that yogis could stop their hearts, see Therese Brosse, "A Psychophysiological Study," *Main Currents in Modern Thought* 4 (1946): 77–84. For a general review of this older literature, see Michael Murphy and Steven Donovan, *The Physical and Psychological Effects of Meditation: A Review of Contemporary Research with a Comprehensive Bibliography, 1931–1996*, 2d ed. (Sausalito, Calif.: Institute of Noetic Sciences,1997), also available online at www.noetic.org/research/medbiblio/index.htm.
76. Gay Luce and Erik Peper, "Mind over Body, Mind over Mind," *New York Times*, September 12, 1971.
77. H. Benson, J. W. Lehmann, M. S. Malhotra, R. F. Goldman, et al., "Body Temperature Changes during the Practice of g Tum-mo Yoga." *Nature* 295, no. 5846 (1982): 234–36. For criticism and Benson's response, see "Letters to the editor" on "Body Temperature Changes During the Practice of g Tum-mo Yoga," *Nature* 298, no. 5872 (1982): 402.
78. The footage is currently housed in the Herbert Benson archive at the Countway Library of Harvard Medical School, where it is available to researchers.
79. The story of the Boston studies, which caused controversy in the Tibetan community, is briefly discussed in Zara Houshmand, Anne Harrington, Clifford Saron, and Richard J. Davidson, "Training the Mind: First Steps in a Cross-Cultural Collaboration in Neuroscientific Research," in *Visions of Compassion: Western Scientists and Tibetan Buddhists Examine Human Nature*, ed. R. J. Davidson and A. Harrington (New York: Oxford University Press, 2002), pp. 3–17.
80. H. Maturana and F. Varela, *Autopoiesis and Cognition: The Realization of the Living* (Boston: D. Reidel, 1980).
81. For more, see the Mind and Life website, www.mindandlife.org. See also Anne Harrington, "Introduction," in *The Dalai Lama at MIT*, ed. A. Harrington and A. Zajonc (Cambridge, Mass.: Harvard University Press, 2006), pp. 3–18.
82. *Gentle Bridges: Conversations with the Dalai Lama on the Sciences of Mind*, ed. J. Hayward and F. J. Varela (Boston: Shambhala Publications, 1992); *Consciousness at the Crossroads: Conversations with the Dalai Lama on Brainscience and Buddhism*, ed. Zara Houshmand, Robert B. Livingston, and B. Allan Wallace (Ithaca, N.Y.: Snow Lion Publications, 1999).
83. The edited proceedings of the 1990 meeting were published as *Healing Emotions: Conversations with the Dalai Lama on Mindfulness, Emotions, and Health*, ed. Daniel Goleman, © Mind and Life Institute, 1997 (republished, Boston: Shambhala Publications, 2003).

84. See Houshmand et al., "Training the Mind," p. 6.
85. Ibid., p. 13.
86. R. J. Davidson and D. J. Goleman, "The Role of Attention in Meditation and Hypnosis: A Psychobiological Perspective on Transformations of Consciousness," *International Journal of Clinical and Experimental Hypnosis* 25, no. 4 (1977): 291–308; R. J. Davidson, D. J. Goleman, and G. E. Schwartz, "Attentional and Affective Concomitants of Meditation: A Cross-Sectional Study," *Journal of Abnormal Psychology* 85, no. 2 (1976): 235–38.
87. For more on these older romantic visions of Tibet, see Peter Bishop, *The Myth of Shangri-La* (Berkeley: University of California Press, 1989). One of the most important figures in the United States involved in positioning the Dalai Lama on a world stage in those years, and in advocating for seeing Tibetan Buddhism as a precious and endangered resource for the entire world, was the Columbia University professor of Tibetan studies Robert Thurman. It was Thurman who had also taken the lead in securing a Nobel Peace Prize for the Dalai Lama in 1989. For more on this in particular, see, among others, Rodger Kamenetz, "Robert Thurman Doesn't *Look* Buddhist," *New York Times*, May 5,1996.
88. Daniel Goleman and Robert A. F. Thurman, eds., *Mind Science: An East-West Dialogue* (Boston: Wisdom Publications, 1991). Participants at this conference included the Dalai Lama, Herbert Benson, Robert Thurman, Howard Gardner, Daniel Goleman, Diana Eck, and participants in the Harvard Mind Science Symposium
89. Henry Dreher, "Recite Your Mantra and Call Me in the Morning," *New York* magazine, May 11, 1998; available online at http://nymag.com/nymetro/health/feature/2664.
90. Joe Loizzo, "The Science of Enlightenment: Meditative Medicine," *Tricycle: The Buddhist Review*, Fall 1998, pp. 88–89.
91. Davidson et al., "Alterations in Brain and Immune Function" (see note 45 above).
92. For more on Ricard's background and motivations, see Barry Boyce, "Two Sciences of Mind," *Shambhala Sun*, September 2005; available online at www.shambhalasun.com.
93. Antoine Lutz, Lawrence L. Greischar, Nancy B. Rawlings, Matthieu Ricard, and Richard J. Davidson, "Long-term Meditators Self-induce High-amplitude Gamma Synchrony during Mental Practice," *Proceedings of the National Academy of Sciences* 101 (November 16, 2004): 16369–73.
94. Stephen S. Hall, "Is Buddhism Good for Your Health?" *New York Times*, September 14, 2003.
95. David Cyranoski, "Neuroscientists See Red over Dalai Lama," news@nature.com, July 25, 2005.
96. Leigh E. Schmidt, "In the Lab With the Dalai Lama," *Chronicle of Higher Education* 52, no. 17 (December 16, 2005): B10.

Conclusion

1. Bill Moyers, ed., *Healing and the Mind* (New York: Doubleday, 1993).
2. Public Affairs Television, Inc., *Healing and the Mind with Bill Moyers: Overview of*

Series Impact (New York: Public Affairs Television Archives, 1994), cited in Navaz Percy Kananjia, "Healing and the Mind: The Politics of Popular Media and the Development of Mind-Body Medicine in 1990s America," senior honors thesis, Department of the History of Science, Harvard University, 2001.

3. Ibid., p. 81; Nancy Matsumoto, "The Burgeoning Art of Healing with the Head: Mind/Body Books Are Big Business," *Los Angeles Times,* September 5, 1994.
4. Daniel Goleman and Joel Gurin, eds., *Mind Body Medicine: How to Use Your Mind for Better Health* (Yonkers, N.Y.: Consumer Reports Books, 1993).
5. I am indebted to Daniel Goleman (email to author, April 6, 2007) for clarifying the link between his book and the Moyers series, and alerting me to the role played by the Fetzer Institute, and especially its chairman, Rob Lehman, in both projects. The role of the Fetzer Institute in catalyzing developments in the modern construction of mind-body medicine more generally would be worthy of further study. This foundation, for example, also helped found and fund the journal *Advances: The Journal of Mind-Body Health*, and it has sponsored several key research initiatives in this area.
6. "Mind/Body Medicine: Hearing Before a Subcommittee of the Committee on Appropriations," United States Senate, One Hundred Fifth Congress, second session, special hearing, September 12, 1998, available online at http://frwebgate.access.gpo.gov/cgi-bin/getdoc.cgi?dbname=105_senate_hearings&docid=f:54619.wais.
7. Cf. here the ideas of Stephen Hilgartner, "The Dominant View of Popularization: Conceptual Problems, Political Uses," *Social Studies of Science* 20, no. 3 (1990): 519–39. Hilgartner criticizes the widespread notion that science communication is always one-way ("downstream"), from a professional scientific "source" to the wider culture. He argues that sometimes science communication also works "upstream," with ideas moving from the wider culture back to professional communities of science.
8. See especially T. J. Jackson Lears, *No Place of Grace* (New York: Pantheon, 1981), p. 6; and "From Salvation to Self-Realization: Advertising and the Therapeutic Roots of the Consumer Culture, 1880–1930," in *The Culture of Consumption: Critical Essays in American History, 1880–1980*, ed. Richard Wightman Fox and T. J. Jackson Lears (New York: Pantheon Books, 1983). Earlier classic texts in this broad historiographic tradition are Philip Rieff, *Freud: The Mind of the Moralist*, 3d ed. (Chicago: University of Chicago Press, 1979), and Christopher Lasch, *The Culture of Narcissism: American Life in an Age of Diminishing Expectations* (New York: W. W. Norton, 1978).
9. Roy M. Anker, *Self-Help and Popular Religion in Early American Culture: An Interpretive Guide to Origins* (Westport, Conn.: Greenwood Press, 1999); Eva S. Moskowitz, *In Therapy We Trust: America's Obsession with Self-Fulfillment* (Baltimore, Md.: Johns Hopkins University Press, 2001); Steven Starker, *The Oracle at the Supermarket: The American Preoccupation with Self-Help Books* (New Brunswick, N.J.: Transaction Books, 2002).
10. These statistics are from Moskowitz, *In Therapy We Trust*, but I took them from a thoughtful essay review that discussed her book: Christopher P. Loss, "Religion and the Therapeutic Ethos in Twentieth-Century American History," *American Studies International* 40, no. 3 (2002): 6–76.

11. For more on the role of World War II in creating this new profession of psychological experts, see Ellen Herman, *The Romance of American Psychology* (Berkeley: University of California Press, 1995).
12. Institute of Medicine, *Research on Stress and Human Health* (Washington, D.C.: National Academy Press, 1981): pp. 2–3.
13. Anne Harrington, ed. (Cambridge, Mass.: Harvard University Press, 1997); Anne Harrington, "Seeing the Placebo Effect: Historical Legacies and New Opportunities," in *The Science of the Placebo*, ed. Harry A. Guess, Arthur Kleinman, John W. Kusek, and Linda W. Engel (London: BMJ Books, 2002); Richard Davidson, Jon Kabat-Zinn, Jessica Schumacher, Melissa Rosenkranz, et al., "Alterations in Brain and Immune Function Produced by Mindfulness Meditation," *Psychosomatic Medicine* 65 (2003): 564–70; Anne Harrington and Arthur Zajonc, eds., *The Dalai Lama at MIT* (Cambridge, Mass.: Harvard University Press, 2006); Anne Harrington, "Finding Qi and Chicanery in China," *Spirituality & Health* (Summer 2001); Anne Harrington, "Eastward Journeys and Western Discontents: China, 'Qi,' and the Challenges of 'Dual Vision,' " in *Science, History, and Social Activism*, ed. Garland E. Allen and Roy M. MacLeod (Boston: Kluwer Academic, 2001).
14. S. E. Mouridsen and S. Nielsen, "Reversible Somatotropin Deficiency (Psychosocial Dwarfism) Presenting as Conduct Disorder and Growth Hormone Deficiency," *Developmental Medicine & Child Neurology* 32, no. 12 (1990): 1093–98.
15. David P. Phillips and Daniel G. Smith, "Postponement of Death Until Symbolically Meaningful Occasions," *Journal of the American Medical Association* 263, no. 14 (1990): 1947–51.
16. P. Cooke, "They Cried Until They Could Not See," *New York Times Magazine*, June 23, 1991.
17. Bruno Klopfer, "Psychological Variables in Human Cancer," *Journal of Projective Techniques* 31, no. 4 (December 1957): 331–40.
18. Virginia A. Hoffman, Anne Harrington, and Howard Fields, "Pain and the Placebo: What We Have Learned," *Perspectives in Biology and Medicine* 48, no. 2 (2005): 248–65.
19. Sidney Cobb and Robert M. Rose, "Hypertension, Peptic Ulcer and Diabetes in Air Traffic Controllers," *Journal of the American Medical Association* 224, no. 4 (1973): 489–92, cited p. 492.
20. Interview with Robert Rose by Anne Harrington, June 5–6, 1997.
21. Misia Landau, "Human Evolution as Narrative," *American Scientist* 72 (1984): 262–68.

Select Bibliography

Ader, Robert. "Presidential Address: Psychosomatic and Psychoimmunologic Research." *Psychosomatic Medicine* 42 (1980): 307–22.

———. "On the Development of Psychoneuroimmunology." *European Journal of Pharmacology* 405 (September 29, 2000): 167–76.

Ader, Robert, ed. *Psychoneuroimmunology*. New York: Academic Press, 1981.

Ader, Robert, and Nicholas Cohen. "Behaviorally Conditioned Immunosuppression." *Psychosomatic Medicine* 37, no. 4 (July–August 1975): 333–40.

Alexander, Franz. "Fundamental Concepts of Psychosomatic Research." *Psychosomatic Medicine* 5 (1943): 205–10.

———. *Psychosomatic Medicine: Its Principles and Applications*. New York: W. W. Norton, 1950.

———. "The Psychosomatic Approach in Medical Therapy." [1954.] In *The Scope of Psychoanalysis, 1921–1961: Selected Papers*. New York: Basic Books, 1961.

Alexander, Franz, and Thomas Morton French. *Psychoanalytic Therapy: Principles and Applications*. New York: Ronald Press, 1946.

Alexander, P. S. "The Demonology of the Dead Sea Scrolls." In *The Dead Sea Scrolls after Fifty Years: A Comprehensive Assessment*, vol. 2, ed. Peter W. Flint and James C. VanderKam. Leiden: E. J. Brill, 1999. Pp. 331–53.

Allen, Garland E., and Roy M. MacLeod, eds. *Science, History, and Social Activism: A Tribute to Everett Mendelsohn*. Boston: Kluwer Academic, 2001.

Amick, Benjamin, Sol Levine, Alvin R. Tarlov, and Diana Chapman Walsh, eds. *Society and Health*. New York: Oxford University Press, 1995.

Angell, Marcia. "Disease as a Reflection of the Psyche." *New England Journal of Medicine* 312, no. 4 (June 13, 1985): 1570–72.

Anonymous. "The Powerful Placebo and the Wizard of Oz." *New England Journal of Medicine* 344 (2001): 1630–32.

Anonymous. "Science Surrenders to the Inexplicable: Jean-Pierre Bely Was Instantly Cured of Multiple Sclerosis." ZENIT news agency, February 11, 1999. Available online at www.zenit.org/english/archive/9902/ZE990211.html.

Antze, Paul, and Michael Lambek, eds. *Tense Past: Cultural Essays in Trauma and Memory*. New York: Routledge, 1996.

Appignanesi, Lisa, and John Forrester. *Freud's Women*. New York: Basic Books, 1992.

Aronowitz, Robert, and Howard M. Spiro. "The Rise and Fall of the Psychosomatic Hypothesis in Ulcerative Colitis." *Journal of Clinical Gastroenterology* 10 (1988): 298–305.

Aronowitz, Robert A. *Making Sense of Illness: Science, Society, and Disease*. New York: Cambridge University Press, 1998.

Auden, W. H. *Collected Poems*. Ed. Edward Mendelsohn. New York: Vintage Books, 1991.

AvRuskin, Tara L. "Neurophysiology and the Curative Possession Trance: The Chinese Case." *Medical Anthropology Quarterly* 2, no. 3 (1988): 286–302.

Babinski, J. "Démembrement de l'hystérie traditionelle: Pithiatisme." *Semaine médicale* 29, no. 1 (January 6, 1909): 3–8.

Bancroft, Mark. *The History & Psychology of Spirit Possession & Exorcism*. EnSpire Press, 1998. Available online at www.enspirepress.com/writings_on_consciousness/spirit_possession_exorcism/spirit_possession_exorcism.html.

Barke, Megan, Rebecca Fribush, and Peter N. Stearns. "Nervous Breakdown in 20th-Century American Culture." *Journal of Social History* 3, no. 3 (2000): 565–84.

Barnes, J. A. "Class and Committees in a Norwegian Island Parish," *Human Relations* 7 (1954): 39–58.

Beard, George M. *American Nervousness, Its Causes and Consequences: A Supplement to Nervous Exhaustion*. New York: Putnam, 1881.

Beasley, Norman. *The Cross and the Crown: The History of Christian Science*. New York: Duell, Sloan, and Pearce, 1952.

Beecher, Henry. "The Powerful Placebo." *Journal of the American Medical Association* 159 (1955): 1602–06.

Benison, Saul, A. Clifford Barger, and Elin L. Wolfe. *Walter B. Cannon: The Life and Times of a Young Scientist*. Cambridge, Mass.: Belknap Press, 1987.

Benson, Herbert. "Yoga for Drug Abuse." *New England Journal of Medicine* 281, no. 20 (1969): 1133.

———. *The Relaxation Response*. [1975.] New York: Avon Books, 1976.

Benson, Herbert, J. F. Beary, and M. P. Carol. "The Relaxation Response." *Psychiatry* 37 (1974): 37–46.

Benson, Herbert, J. W. Lehmann, M. S. Malhotra, R. F. Goldman, et al. "Body Temperature Changes during the Practice of g Tum-mo Yoga (Matters Arising)." *Nature* 298, no. 587 (1982): 402.

Benson, Herbert, David Shapiro, Bernard Tursky, and Gary E. Schwartz. "Decreased Systolic Blood Pressure Through Operant Conditioning Techniques in Patients with Essential Hypertension." *Science* 173, no. 998 (1971): 740–42.

Berczi, Istvan. "Stress and Disease: The Contributions of Hans Selye to Neuroimmune Biology. A Personal Reminiscence." Available online at http://home.cc.umanitoba.ca/~berczii/page2.htm. Reprinted from Istvan Berczi and Judith Szélenyi, eds., *Advances in Psychoneuroimmunology*. New York: Plenum, 1994.

Berkman, Lisa F., T. Glass, I. Brisette, and T. E. Seeman. "From Social Integration to Health: Durkheim in the New Millennium." *Social Science and Medicine* 51 (2000): 843–57.

Berkman, Lisa F., and S. Leonard Syme. "Social Networks, Host Resistance, and Mortality: A Nine-Year Follow-up Study of Alameda County Residents." *American Journal of Epidemiology* 109, no. 2 (1979): 186–204.

Bernheim, Hippolyte. *De la suggestion dans l'état hypnotique et dans l'état de veille*. Paris: Octave Doin, 1884.

———. *Suggestive Therapeutics: A Treatise on the Nature and Uses of Hypnotism*. Trans. Christian A. Herter. New York: G. P. Putnam's, 1889. Originally published as *De la suggestion et de ses applications thérapeutique*, 2nd ed. Paris: Octave Doin, 1886.

Bernikow, Louise. "Alone: Yearning for Companionship in America." *New York Times Magazine*, August 15, 1982.

———. *Alone in America: The Search for Companionship*. New York: Harper & Row, 1986.

Bertrand, Alexandre. *Du Magnétisme Animal en France*. Paris: J. B. Baillière, 1826.

Besnard, P. "Anomie." In *International Encyclopedia of the Social and Behavioral Sciences*. Editors in chief Neil J. Smelser and Paul B. Baltes. New York: Elsevier, 2001. Pp. 510–13.

Bishop, Peter. *The Myth of Shangri-La: Tibet, Travel Writing, and the Western Creation of a Sacred Landscape*. Berkeley: University of California Press, 1989.

Blumberg, E. M., P. M. West, and F. W. Ellis. "A Possible Relationship between Psychological Factors and Human Cancer." *Psychosomatic Medicine* 16 (1954): 277–86.

Bogacz, Ted. "War Neurosis and Cultural Change in England, 1914–22: The Work of the War Office Committee of Enquiry into 'Shell-Shock.' " *Journal of Contemporary History* 24, no. 2 (April 1989): 227–56.

Booker, Christopher. *The Seven Basic Plots: Why We Tell Stories*. London: Continuum, 2004.

Borch-Jacobsen, Mikkel. *Remembering Anna O: A Century of Mystification*. Trans. Kirby Olson. New York: Routledge, 1996.

Bothe, Detlef. *Neue Deutsche Heilkunde 1933–1945*. Husum: Matthiesen Verlag, 1991.

Bourguignon, Erika. "World Distribution and Patterns of Possession States." In *Trance and Possession States: Proceedings Second Annual Conference, R. M. Bucke Memorial Society, 4–6 March 1966, Montreal*. Ed. Raymond Prince. Montreal: R. M. Bucke Memorial Society, 1968.

———. *Possession*. San Francisco: Chandler & Sharp, 1976.

Bourguignon, Erika, ed. *Religion, Altered States of Consciousness, and Social Change*. Columbus: Ohio State University Press, 1973.

Bowlby, John. *Maternal Care and Mental Health*. Geneva: World Health Organization, 1951.

Braden, Charles. *Spirits in Rebellion: The Rise and Development of New Thought*. Dallas, Tex.: Southern Methodist University Press, 1963.

Brady, Joseph V. "Ulcers in 'Executive' Monkeys." *Scientific American* 199 (1958): 95–100.

Braid, James. *Neurypnology, or, the Rationale of Nervous Sleep, Considered in Relation with Animal Magnetism.* [1843.] London: Churchill, 1848.

Breuer, Josef, and Sigmund Freud. *Studies on Hysteria.* [1895.] Trans. and ed. James Strachey. A reprint of vol. 2 of *The Standard Edition of the Complete Psychological Works of Sigmund Freud* (Hogarth Press, 1955). New York: Basic Books, 1957.

Bridge, T. Peter, Allan F. Mirsky, and Frederick K. Goodwin, eds. *Psychological, Neuropsychiatric, and Substance Abuse Aspects of AIDS.* New York: Raven Press, 1988.

Brisset, Dennis, and Charles Edgley, eds. *Life as Theater: A Dramaturgical Sourcebook.* New York: Aldine de Gruyter, 1990.

Brody, Howard, and Daralyn Brody. *The Placebo Response: How You Can Release the Body's Inner Pharmacy for Better Health.* New York: Cliff Street Books, 2000.

Brosse, Therese. "A Psychophysiological Study." *Main Currents in Modern Thought* 4 (1946): 77–84.

Brown, Steven D. "The Life of Stress: Seeing and Saying Dysphoria." Ph.D. diss., University of Reading, 1998.

Brown, Theodore M. "Descartes, Dualism, and Psychosomatic Medicine." In *The Anatomy of Madness: Essays in the History of Psychiatry*, vol. 1, ed. W. F. Bynum, Roy Porter, and Michael Shepherd. New York: Tavistock, 1985.

———. "Alan Gregg and the Rockefeller Foundation's Support of Franz Alexander's Psychosomatic Research." *Bulletin of the History of Medicine* 61, no. 2 (1987); 155–58.

———. "The Rise and Fall of American Psychosomatic Medicine." Paper presented to New York Academy of Medicine, New York, November 29, 2000. Available onlne at www.human-nature.com/free-associations/riseandfall.html.

Bruckbauer, Elizabeth, and Sandra E. Ward. "Positive Mental Attitude and Health: What the Public Believes." *Image: Journal of Nursing Scholarship* 25 (1993): 311–15.

Bruhn, J. G., B. Phillips, and S. Wolf. "Social Readjustment and Illness Patterns: Comparison between First, Second, and Third Generation Italian-Americans Living in the Same Community." *Psychosomatic Medicine* 16 (1972): 387–94.

Bruhn, John G., and Stewart Wolf. *The Roseto Story: An Anatomy of Health.* Norman: University of Oklahoma Press, 1979.

Buell, Lawrence. *Literary Transcendentalism: Style and Vision in the American Renaissance.* Ithaca, N.Y.: Cornell University Press, 1973.

Buranelli, Vincent. *The Wizard from Vienna: Franz Anton Mesmer.* London: Peter Owen, 1975.

Burchfield, S. R. "The Stress Response: A New Perspective." *Psychosomatic Medicine* 41 (1979): 661–72.

Bynum, Caroline Walker. *Holy Feast and Holy Fast: The Religious Significance of Food to Medieval Women.* Berkeley: University of California Press, 1987.

———. "Why All the Fuss about the Body?" *Critical Inquiry* 22 (1995): 1–33.

Cannon, Walter B. *The Mechanical Factors of Digestion.* New York: Longmans, Green & Co., 1911.

———. *Bodily Changes in Pain, Hunger, Fear, and Rage: An Account of Recent Researches into the Function of Emotional Excitement.* 2d ed. New York: D. Appleton, 1929.

———. "The Role of Emotions in Disease." *Annals of Internal Medicine* 9, no. 11 (May 1936): 1453–65.

———. *The Way of an Investigator: A Scientist's Experiences in Medical Research.* New York: W. W. Norton, 1945.

Cannon, Walter B., and Daniel de la Paz. "Emotional Stimulation of Adrenal Secretion." *American Journal of Physiology* 28 (1911): 64–70.

Cannon, Walter B., and Arturo Rosenblueth. *Autonomic Neuro-effector Systems*. New York: Macmillan, 1937.

Carus, Carl Gustav. *Psyche, zur Entwicklungsgeschichte der Seele*. Pforzheim: Flammer and Hoffmann, 1846.

Cassel, John. "Psychosocial Processes and Stress: Theoretical Formulations." *International Journal of Health Services* 4 (1974): 471–82.

———. "The Contribution of the Social Environment to Host Resistance." *American Journal of Epidemiology* 104 (1976): 107–23.

Charcot, J. M. "Sur les divers états nerveux déterminés par l'hypnotisation chex lex hystériques." *Comptes Rendues de l'Academie des Sciences* 94 (1882): 403–05.

———. "La foi qui guérit." *Revue Hebdomadaire* 5 (1892): 112–32. Published in English as "The Faith Cure," *New Review* 11 (1892): 244–62.

———. *Charcot the Clinician: The Tuesday Lessons: Excerpts from Nine Case Presentations on General Neurology Delivered at the Salpêtrière Hospital in 1887–88 by Jean-Martin Charcot*. Trans. with commentary by Christopher G. Goetz. New York: Raven Press, 1987.

Cloud, David W. "Norman Vincent Peale: Apostle of Self-Esteem." O *Timothy* 11, no. 2 (1994). Available online at www.wayoflife.org/fbns/fbns473.html.

Cobb, Sidney. "Social Support as a Moderator of Life Stress." *Psychosomatic Medicine* 38 (1976): 300–314.

Cobb, Sidney, and Robert M. Rose. "Hypertension, Peptic Ulcer and Diabetes in Air Traffic Controllers." *Journal of the American Medical Association* 224, no. 4 (1973): 489–92.

Cohen, Sanford I. "Voodoo Death, the Stress Response, and AIDS." In *Psychological, Neuropsychiatric, and Substance Abuse Aspects of AIDS*, ed. T. Peter Bridge, Allan F. Mirsky, and Frederick K. Goodwin. New York: Raven Press, 1988.

Cohen, Sheldon, and S. Leonard Syme, eds. *Social Support and Health*. London: Academic Press, 1985.

Comfort, Kathleen Ann. "Divine Images of Hysteria in Emile Zola's *Lourdes*." *Nineteenth Century French Studies* 30, no. 3 and 4 (2002): 330–46.

Condie, Lana. "The Practice of Exorcism and the Challenge to Clerical Authority." *Access: History* 3, no. 1 (2000): 93–103.

Cooper, T., T. Detre, and S. M. Weiss. The Review Panel on Coronary-Prone Behavior and Heart Disease. "Coronary-prone Behavior and Coronary Heart Disease: A Critical Review." *Circulation* 63 (1981): 1199–1215.

Coppens, A., and M. Metcalfe. "Cancer and Extroversion." *British Medical Journal* 236 (1963): 18–19.

Cousins, Norman. "Anatomy of an Illness (as Perceived by the Patient)." *New England Journal of Medicine* 295, no. 26 (December 23, 1976): 1458–63.

———. *Anatomy of an Illness as Perceived by the Patient: Reflections on Healing and Regeneration*. New York: W. W. Norton, 1979.

Cowley, Geoffrey, with Anne Underwood. "Is Love the Best Drug?" *Newsweek*, March 16, 1998.

Cox, Harvey. *Turning East: The Promise and Peril of the New Orientalism*. New York: Simon & Schuster, 1977.

Crabtree, Adam. *From Mesmer to Freud*. New Haven, Conn.: Yale University Press, 1993.

Cromie, William J. "Meditation Changes Temperatures: Mind Controls Body in Extreme Experiments." *Harvard University Gazette*, April 18, 2002. Available online at www.hno.harvard.edu/gazette/2002/04.18/09-tummo.html.

Darnton, Robert. *Mesmerism and the End of the Enlightenment in France*. Cambridge, Mass.: Harvard University Press, 1968.

Davidson, Richard J., and Anne Harrington, eds. *Visions of Compassion: Western Scientists and Tibetan Buddhists Examine Human Nature*. New York: Oxford University Press, 2002.

Davidson, Richard J., Jon Kabat-Zinn, Jessica Schumacher, Melissa Rosenkranz, Daniel Muller, Saki F. Santorelli, Ferris Urbanowski, Anne Harrington, Katherine Bonus, and John F. Sheridan. "Alterations in Brain and Immune Function Produced by Mindfulness Meditation." *Psychosomatic Medicine* 65 (2003): 564–70.

De Marneffe, Daphne. "Looking and Listening: The Construction of Clinical Knowledge in Charcot and Freud." *Signs* 17, no. 1 (1991): 91–112.

Dembroski, T. M., J. M. MacDougal, R. B. Williams, et al. "Components of Type A, Hostility, and Anger in Relationship to Angiographic Findings." *Psychosomatic Medicine* 47 (1985): 219–24.

Deutsch, Felix. "On the Formation of the Conversion Symptom." In *On the Mysterious Leap from the Mind to the Body: A Workshop Study on the Theory of Conversion*, ed. Felix Deutsch. New York: International Universities Press, 1959.

Didi-Huberman, Georges. *Invention de l'hystérie: Charcot et l'Iconographie photographique de la Salpêtrière*. Paris: Macula, 1992.

———. *Invention of Hysteria: Charcot and the Photographic Iconography of the Salpêtrière*. Trans. Alisa Hartz. Cambridge, Mass.: MIT Press, 2003.

Dienstfrey, Harris. *Where the Mind Meets the Body*. New York: HarperCollins, 1991.

Dohrenwend, Barbara Snell, and Bruce P. Dohrenwend, eds. *Stressful Life Events: Their Nature and Effects*. New York: John Wiley & Sons, 1974.

Domash, Lawrence J. "The Transcendental Meditation Technique and Quantum Physics." In *Scientific Research on Maharishi's Transcendental Meditation and TM-Sidhi Program: Collected Papers*, vol. 1, 2d ed., ed. David W. Orme-Johnson and John T. Farrow. West Germany: Maharishi European University Press, 1977. Pp. 652–70.

Duden, Barbara. *The Woman Beneath the Skin: A Doctor's Patients in Eighteenth-Century Germany*. Cambridge, Mass.: Harvard University Press, 1991.

Dunbar, H. Flanders. "Character and Symptom Formation: Some Preliminary Notes with Special Reference to Patients with Hypertension, Rheumatic, and Coronary Disease." *Psychoanalytic Quarterly* 8 (1939): 18–47.

———. "Medical Aspects of Accidents and Mistakes in the Industrial Army and in the Armed Forces." *War and Medicine* 4 (1943): 161–75.

Dunbar, H. Flanders, and Leon Brody. "Basic Aspects and Applications of the Psychology of Safety." New York: Center for Safety Education, New York University, 1959.

Dunn, A. J. "Psychoneuroimmunology: Introduction and General Perspectives." In *Stress, the Immune System, and Psychiatry*, ed. B. Leonard and K. Miller. New York: John Wiley & Sons, 1995.

Durdin, Peggy. "Medicine in China: A Revealing Story." *New York Times*, February 28, 1960.

Durkheim, Émile. *De la division du travail social: Étude sur l'organisation des sociétes supérieures*. Paris: F. Alcan, 1893. Published in English as *The Division of Labor in Society*. Trans. George Simpson. New York: Free Press of Glencoe, 1933.

———. *Le Suicide: Étude de sociologie*. Paris: F. Alcan, 1897. Published in English as *Suicide: A Study in Sociology*. Trans. John A. Spaulding and George Simpson. Glencoe, Ill.: Free Press, 1951.

Ebon, M. *Maharishi: The Guru*. London: Pearl, 1967.

Eck, Diana. "New Age Hinduism in America." In *Conflicting Images: India and the United States*, ed. Nathan Glazer and Sulochana Glazer. Glenndale, Md.: Riverdale, 1993.

Eddy, Mary Baker. "Retrospection and Introspection." Boston: Trustees under the Will of Mary Baker G. Eddy, 1891. Available online at www.christianscience.org/Prose%20Works/RetroIntro.html.

———. *Science and Health with Key to the Scriptures*. Boston: Christian Science Board of Directors, 1994.

Edinger, Dora. *Bertha Pappenheim: Freud's Anna O.* Highland Park, Ill.: Congregation Solel, 1968.

Egolf, B., J. Lasker, S. Wolf, and L. Potvin. "The Roseto Effect: A 50-Year Comparison of Mortality Rates." *American Journal of Public Health* 82, no. 8 (August 1992): 1089–92.

Eisenberg, David. "Energy Medicine in China: Defining a Research Strategy which Embraces the Criticism of Skeptical Colleagues." *Noetic Sciences Review* 14 (Spring 1990): 7.

———. "The Mystery of Chi: Medicine in a Mind/Body Culture and Another Way of Seeing." In *Healing and the Mind*, ed. Bill Moyers. New York: Doubleday, 1993. Pp. 51–314.

Eisenberg, David M., R. C. Kessler, C. Fsterm, F. E. Norlock, D. R. Calkins, and T. L. Delbanco. "Unconventional Medicine in the United States: Prevalence, Costs, and Patterns of Use." *New England Journal of Medicine* 328, no. 4 (1993): 246–52.

Elias, Norbert. *The History of Manners*. Trans. Edmund Jephcott. New York: Pantheon Books, 1982.

———. *Power and Civility*. Trans. Edmund Jephcott. New York: Pantheon Books, 1982.

Ellenberger, Henri. *The Discovery of the Unconscious: The History and Evolution of Dynamic Psychiatry*. New York: Basic Books, 1970.

———. "The Story of Anna O.: A Critical Review with New Data." *Journal of the History of the Behavioral Sciences* 8 (July 1972): 267–79.

Emerson, Ralph Waldo. "The Transcendentalist." [1842.] In *Nature: Addresses and Lectures* [1849], reprinted in *Essays & Poems*. New York: Library of America, 1996.

Engel, George L. "Sudden and Rapid Death during Psychological Stress: Folklore or Folk Wisdom?" *Annals of Internal Medicine* 74 (1971): 771–82.

Farrar, W. L. "Evidence for the Common Expression of Neuroendocrine Hormones and Cytokines in the Immune and Central Nervous Systems." *Brain, Behavior, Immunity* 2 (1988): 322–27.

Ferenczi, S. "The Principle of Relaxation and Neocatharsis." [1930.] In S. Ferenczi, *Final Contributions to the Problems and Methods of Psycho-Analysis*, ed. M. Balint, trans. E. Mosbacher et al. London: Hogarth Press, 1955.

———. "Confusion of Tongues Between Adults and the Child (the Language of Tenderness and of Passion)." [1933.] In S. Ferenczi, *Final Contributions to the Problems and*

Methods of Psycho-Analysis, ed. M. Balint, trans. E. Mosbacher et al. London: Hogarth Press, 1955.

———. *Final Contributions to the Problems and Methods of Psycho-Analysis*. Ed. M. Balint, trans. E. Mosbacher et al. London: Hogarth Press, 1955.

Fiske, Edward B. "Thousands Finding Meditation Eases Stress of Living." *New York Times*, December 11, 1972.

Flanagan, Cara. *Early Socialisation: Sociability and Attachment*. New York: Routledge, 1999.

Flint, Peter W., and James C. VanderKam, eds. *The Dead Sea Scrolls after Fifty Years: A Comprehensive Assessment*. 2 vol. Leiden: E. J. Brill, 1998–1999.

Fox, Richard Wightman, and T. J. Jackson Lears, eds. *The Culture of Consumption: Critical Essays in American History, 1880–1980*. New York: Pantheon Books, 1983.

Frank, Arthur W. *The Wounded Storyteller: Body, Illness, and Ethics*. Chicago: University of Chicago Press, 1995.

Franklin, Benjamin, et al. *Testing the Claims of Mesmerism*. An English translation of the 1784 report by Benjamin Franklin et al. on Mesmerism. Trans. Charles and Danielle Salas, introduction by Michael Shermer. *Skeptic* 4, no. 3 (1996): 66–83.

Fraser, Caroline. "Suffering Children and the Christian Science Church." *The Atlantic Monthly* 264, no. 4 (April 1995): 105–20.

Freeman, L. *The Story of Anna* O. New York: Walker, 1972.

Freud, Sigmund. "Preface to the Translation of Bernheim's *Suggestion*." [1888.] In *The Standard Edition of the Complete Psychological Works of Sigmund Freud*, vol. 1, 1886–1899, *Pre-psychoanalytic Publications and Unpublished Drafts*. Trans. James Strachey. New York: Vintage (Hogarth Press), 2000.

———. "The Neuro-psychoses of Defense." [1894.] In *The Standard Edition of the Complete Psychological Works of Sigmund Freud*, vol. 3, 1893–1899, *Early Psycho-analytic Publications*. Trans. James Strachey. New York: Vintage (Hogarth Press), 2000.

———. "The Aetiology of Hysteria." [1896.] In *The Standard Edition of the Complete Psychological Works of Sigmund Freud*, vol. 3: 1893–1899, *Early Psycho-analytic Publications*. Trans. James Strachey. New York: Vintage (Hogarth Press), 2000.

———. "Preface to the Second German Edition of Bernheim's *Suggestion*." [1896.] In *The Standard Edition of the Complete Psychological Works of Sigmund Freud*, vol. 1, 1886–1899, *Pre-psychoanalytic Publications and Unpublished Drafts*. Trans. James Strachey. New York: Vintage (Hogarth Press), 2000.

———. "My Views on the Part Played by Sexuality in the Aetiology of the Neuroses." [1905.] In *The Standard Edition of the Complete Psychological Works of Sigmund Freud*, vol. 7, 1901–1905, *A Case of Hysteria, Three Essays on Sexuality, and Other Works*. Trans. James Strachey. New York: Vintage (Hogarth Press), 2000.

———. "Three Essays on the Theory of Sexuality." [1905.] In *The Standard Edition of the Complete Psychological Works of Sigmund Freud*, vol. 7: 1901–1905, *A Case of Hysteria, Three Essays on Sexuality, and Other Works*. Trans. James Strachey. New York: Vintage (Hogarth Press), 2000.

———. "Introductory Lectures on Psycho-analysis, Part 3." [1916–1917.] In *The Standard Edition of the Complete Psychological Works of Sigmund Freud*, vol. 16: 1916–1917, *Introductory Lectures on Psycho-analysis, Part 3*. Trans. James Strachey. New York: Vintage (Hogarth Press), 2000.

———. "Introduction, Psycho-analysis and the War Neuroses" [1919.] In *The Standard*

Edition of the Complete Psychological Works of Sigmund Freud, vol. 17: 1917–1919, *An Infantile Neurosis and Other Works*. Trans. James Strachey. New York: Vintage (Hogarth Press), 2000.

———. "Memorandum on the Electrical Treatment of War Neurotics." [1920.] In *The Standard Edition of the Complete Psychological Works of Sigmund Freud*, vol. 17: 1917–1919, *An Infantile Neurosis and Other Works*. Trans. James Strachey. New York: Vintage (Hogarth Press), 2000.

———. *Group Psychology and the Analysis of the Ego*. [1921.] Trans. James Strachey. New York: W. W. Norton, 1951.

———. *Letters of Sigmund Freud.* Ed. Ernst L. Freud. Trans. Tani Stern and James Stern. New York: Basic Books, 1960.

———. "Papers on Hypnotism and Suggestion: Editor's Introduction." In *The Standard Edition of the Complete Psychological Works of Sigmund Freud*, vol. 1: 1886–1899, *Pre-psychoanalytic Publications and Unpublished Drafts*. Trans. James Strachey. New York: Vintage (Hogarth Press), 2000.

Friedman, Meyer, and Ray H. Rosenman. "Association of Specific Overt Behavior Pattern with Blood and Cardiovascular Findings." *Journal of the American Medical Association* 169 (1959): 1286–96.

———. *Type A Behavior and Your Heart*. New York: Knopf, 1974.

Friedman, Meyer, C. E. Thoresen, J. J. Gill, et al. "Alteration of Type A Behavior and Its Effect on Cardiac Recurrences in Post-Myocardial Infarction Patients: Summary Results of the Recurrent Coronary Prevention Project." *American Heart Journal* 112 (1986): 653–62.

Friedman, Meyer, and D. Ulmer. *Treating Type "A" Behaviour and Your Heart*. London: Guild, 1985.

Frothingham, Octavious Brooks. *Transcendentalism in New England: A History.* [1876.] Philadelphia: University of Pennsylvania Press, 1972.

Fruehauf, Heiner. "Science, Politics, and the Making of TCM: Chinese Medicine in Crisis." *Journal of Chinese Medicine* 61 (October, 1999): 1–9.

Furst, Lillian R. *Just Talk: Narratives of Psychotherapy*. Lexington: University Press of Kentucky, 1999.

Gauld, Alan. *A History of Hypnotism*. New York: Cambridge University Press, 1992.

George, Carol V. R. *God's Salesman: Norman Vincent Peale and the Power of Positive Thinking*. New York: Oxford University Press, 1993.

Gergen, Kenneth J. "Narrative, Moral Identity and Historical Consciousness: A Social Constructionist Account." 1997. Available online at www.swarthmore.edu/SocSci/kgergen1/web/page.phtml?id=manu38st=manuscriptsllit=1.

Giddens, Anthony. "Beyond Chaos and Dogma. An interview with George Soros, Financier." *The Statesman*, October 31, 1997.

Giovacchini, P. L., and H. Muslin. "Ego Equilibrium and Cancer of the Breast." *Psychosomatic Medicine* 27 (1965): 524.

Glaser, Ronald, Janice K. Kiecolt-Glaser, R. H. Bonneau, et al. "Stress-Induced Modulation of the Immune Response to Recombinant Hepatitis B Vaccine." *Psychosomatic Medicine* 54, no. 1 (1992): 22–29.

Glaser, Ronald, Janice K. Kiecolt-Glaser, W. B. Malarkey, and J. F. Sheridan. "The Influence of Psychological Stress on the Immune Response to Vaccines." *Annals of the New York Academy of Sciences* 840 (1998): 649–55.

Glazer, Nathan, and Sulochana Glazer, eds. *Conflicting Images: India and the United States*. Glenndale, Md.: Riverdale, 1993.

Goffman, Erving. *The Presentation of Self in Everyday Life*. Garden City, N.Y.: Doubleday, 1959.

Goldstein, Jan. "The Hysteria Diagnosis and the Politics of Anticlericalism in Late Nineteenth-Century France." *Journal of Modern History* 54 (June 1982): 209–39.

———. "Enthusiasm or Imagination? Eighteenth-Century Smear Words in Comparative National Context." In *Enthusiasm and Enlightenment in Europe, 1650–1850*, ed. Lawrence E. Klein and Anthony J. La Vopa. San Marino, Calif.: Huntington Library, 1998. Originally published in *Huntington Library Quarterly* 60 (1998): 29–49.

Goleman, Daniel. "Finding Happiness: Cajole Your Brain to Lean to the Left." *New York Times*, February 4, 2003.

Goleman, Daniel, ed. *Healing Emotions: Conversations with the Dalai Lama on Mindfulness, Emotions, and Health*. Boston: Shambhala Publications, 2003.

Goleman, Daniel, and Joel Gurin. *Mind Body Medicine: How to Use Your Mind for Better Health*. Yonkers, N.Y.: Consumer Reports Books, 1993.

Goleman, Daniel, and Robert A. F. Thurman, eds. *MindScience: An East-West Dialogue*. Boston: Wisdom Publications, 1991.

Goodare, Heather. "A Scientific Pioneer of Cancer Groups." *Advances: The Journal of Mind-Body Health* 10, no. 4 (1994): 71–73.

Goodwin, Pamela J., Molyn Leszcz, Marguerite Ennis, Jan Koopmans, Leslie Vincent, Helaine Guther, Elaine Drysdale, Marilyn Hundleby, Harvey M. Chochinov, Margaret Navarro, Michael Speca, and Jonathan Hunter. "The Effect of Group Psychosocial Support on Survival in Metastatic Breast Cancer." *New England Journal of Medicine* 345, no. 24 (2001): 1719–26.

Greenberg, Joel. "The Americanization of Roseto." *Science News* 113, no. 23 (June 10, 1978): 378–80.

Greer, Steven, T. Morris, and K. W. Pettingale. "Psychological Response to Breast Cancer: Effect on Outcome." *Lancet* 2 (1979): 785–87.

Grinker, Roy R., and John P. Spiegel. *Men under Stress*. Philadelphia: Blackison, 1945.

Groddeck, Georg W. *The Book of the It*. [1923.] New York: International Universities Press, 1976. Originally published as *Das Buch vom Es: Psychoanalytische Briefe an eine Freundin*. Vienna: Internationaler Psychoanalytischer Verlag, 1923.

———. *The Meaning of Illness*. [1925.] Trans. Lore Schact. London: Hogarth Press, 1977.

———. "Psychical Treatment of Organic Disease." *British Journal of Medical Psychology* 9 (1929): 179–86.

Guess, Harry A., Arthur Kleinman, John W. Kusek, and Linda W. Engel, eds. *The Science of the Placebo: Toward an Interdisciplinary Research Agenda*. London: BMJ Books, 2002.

Hackett, T. P. "The Psychiatrist: In the Mainstream or on the Banks of Medicine?" *American Journal of Psychiatry* 134 (1977): 432–44.

Hacking, Ian. "The Looping Effect of Human Kinds." In *Causal Cognition: A Multidisciplinary Debate*, ed. Dan Sperber, David Premack, and Ann James Premack. New York: Oxford University Press, 1995. Pp. 351–83.

———. *Rewriting the Soul: Multiple Personality and the Sciences of Memory*. Princeton, N.J.: Princeton University Press, 1995.

Hahn, Peter, ed. *Psychosomatische Medizin: Wege der Forschung*. Darmstadt: Wissenschaftliche Buchgesellschaft, 1985.

Hall, Stephen S. "Is Buddhism Good for Your Health?" *New York Times*, September 14, 2003.

Handfield-Jones, R. P. C. "A Bottle of Medicine from the Doctor." *Lancet* 265 (October 17, 1953): 823–25.

Hansen, Uffe. *Psykoanalysens fortraengte fortid: hypnotisøren Carl Hansen og Sigmund Freud*. Copenhagen: Akademisk Forlag, 1991.

Harrington, Anne. "Metals and Magnets in Medicine: Hysteria, Hypnosis, and Medical Culture in *fin-de-siècle* Paris." *Psychological Medicine* 18, no. 1 (February 1988): 21–38.

———. "Hysteria, Hypnosis, and the Lure of the Invisible: The Rise of Neo-Mesmerism in *fin-de-siècle* French Psychiatry." In *The Anatomy of Madness: Essays in the History of Psychiatry*, Vol. 3, ed. W. F. Bynum, Roy Porter, and Michael Shepherd. New York: Tavistock, 1989.

———. *Reenchanted Science: Holism in German Culture from Wilhelm II to Hitler*. Princeton, N.J.: Princeton University Press, 1996.

———. "Seeing the Placebo Effect: Historical Legacies and New Opportunities." In *The Science of the Placebo: Toward an Interdisciplinary Research Agenda*, ed. Harry A. Guess, Arthur Kleinman, John W. Kusek, and Linda W. Engel. London: BMJ Books, 2002.

———. "Finding Qi and Chicanery in China." *Spirituality & Health*, Summer 2001.

———. "Eastward Journeys and Western Discontents: China, 'Qi,' and the Challenges of 'Dual Vision.' " In *Science, History, and Social Activism: A Tribute to Everett Mendelsohn*, ed. Garland E. Allen and Roy M. MacLeod. Boston: Kluwer Academic, 2001.

Harrington, Anne, ed. *The Placebo Effect: An Interdisciplinary Exploration*. Cambridge, Mass.: Harvard University Press, 1997.

Harrington, Anne, and Arthur Zajonc, eds. *The Dalai Lama at MIT*. Cambridge, Mass.: Harvard University Press, 2006.

Harris, Mark. "The Ties That Bind are the Ties That Heal." *Vegetarian Times* 249 (August 1997): 62–67.

Harris, Ruth. *Murders and Madness: Medicine, Law, and Society in the Fin de Siècle*. New York: Oxford University Press, 1989.

———. *Lourdes: Body and Spirit in the Secular Age*. New York: Viking, 1999.

Harris, T. George. "Heart and Soul: Out on the Cutting Edge, Hard-nosed Researchers Study the Tie between Healthy Emotions and Stout Hearts." *Psychology Today*, January–February 1989. Available online at http://findarticles.com/p/articles/mi_m1175/is_n1_v23/ai.7049288.

Haug, Alfred. *Die Reichsarbeitsgemeinschaft für eine Neue Deutsche Heilkunde (1935–36). Ein Beitrag zur Verhältnis von Schulmedizin, Naturheilkunde und Nationalsozialismus*. Husum: Matthiesen Verlag, 1985.

Heilbron, John L. "The Earliest Missionaries of the Copenhagen Spirit." *Revue d'histoire des sciences* 38, nos. 3 and 4 (1985): 195–230.

Helman, C. G. "Heart Disease and the Cultural Construction of Time: The Type A Behavior Pattern as a Western Culture-bound Syndrome." *Social Science and Medicine* 25, no. 9 (1987): 969–79.

Henig, Robin M. *People's Health: A Memoir of Public Health and Its Evolution at Harvard*. Washington, D.C.: National Academy Press, 1996.

Herman, Ellen. *The Romance of American Psychology*. Berkeley: University of California Press, 1995.

Herman, Judith Lewis. *Trauma and Recovery: The Aftermath of Violence—From Domestic Abuse to Political Terror*. [1989.] New York: Basic Books, 1997.

Hilgartner, Stephen. "The Dominant View of Popularization: Conceptual Problems, Political Uses." *Social Studies of Science* 20, no. 3 (1990): 519–39.

"Hinduism in New York: A Growing Religion." *New York Times*, November 2, 1967.

Hinkle, Lawrence E., Jr. "The Concept of 'Stress' in the Biological and Social Sciences." *Science, Medicine, and Man* 1 (1973): 31–48.

———. "The Effects of Exposure to Culture Change, Social Change, and Changes in Interpersonal Relationships on Health." In *Stressful Life Events: Their Nature and Effects*, ed. Barbara Snell Dohrenwend and Bruce P. Dohrenwend. New York: John Wiley & Sons, 1974.

Hoagland, Hudson. "Adventures in Biological Engineering." *Science* 100 (1944): 63–64.

Hoffman, Virginia A., Anne Harrington, and Howard Field. "Pain and the Placebo: What We Have Learned." *Perspectives in Biology and Medicine* 48, no. 2 (2005): 248–65.

Holden, Constance. "Cousins' Account of Self-Cure Rapped." *Science* 214 (November 1981): 892.

Holland, Jimmie C. "History of Psycho-Oncology: Overcoming Attitudinal and Conceptual Barriers." *Psychosomatic Medicine* 64 (2002): 206–21.

Holmes, Thomas H., and M. Masuda. "Life Change and Illness Susceptibility." In *Stressful Life Events: Their Nature and Effects*, ed. Barbara Snell Dohrenwend and Bruce P. Dohrenwend. New York: John Wiley & Sons, 1974.

Holmes, Thomas H., and Richard H. Rahe. "The Social Readjustment Rating Scale." *Journal of Psychosomatic Research* 11 (1967): 213–18.

Hoskin, Keith. "The Lawful Idea of Accountability": Inscribing People into the Measurement of Objects." In *Accountability: Power, Ethos and the Technologies of Managing*, ed. Rolland Munro and Jan Mouritsen. London: International Thomson Business Press, 1996. Pp. 265–82.

House, James S., K. R. Landis, et al. "Social Relationships and Health." *Science* 241, no. 4865 (1988): 540–45.

House, James S., D. Umberson, and K. R. Landis. "Structures and Processes of Social Support." *American Review of Sociology* 14 (1988): 293–318.

Hrobjartsson, Asbjorn, and Peter C. Gotzsche. "Is the Placebo Powerless? An Analysis of Clinical Trials Comparing Placebo with No Treatment." *New England Journal of Medicine* 344, no. 21 (2001): 1594–1602.

Hunter, Kathryn Montgomery. *Doctors' Stories: The Narrative Structure of Medical Knowledge*. Princeton, N.J.: Princeton University Press, 1991.

James, William. *The Varieties of Religious Experience*. [1902.] New York: Wayne Proudfoot (Barnes and Noble Classics), 2004.

Janet, Pierre. "L'amnésie continue." *Revue Generale des Sciences* 4 (1893): 167–79.

———. *Psychological Healing*. 2 vol. New York: Macmillan, 1925. Originally published as *Les médications psychologiques*. Paris: Félix Alcan, 1919.

Jenkins, Emily. "*Trilby*: Fads, Photographers, and 'Over-Perfect Feet.' " *Book History* 1, no. 1 (1998): 221–67.

Jilek, Wolfgang. "Altered States of Consciousness in Northern American Indian Ceremonies." *Ethos* 10, no. 4 (Winter 1982): 326–42.

Jinpa, Thupten. "Science as an Ally or a Rival Philosophy? Tibetan Buddhist Thinkers' Engagement with Modern Science." In *Buddhism and Science: Breaking New Ground*, ed. Alan B. Wallace. New York: Columbia University Press, 2003.

Jussieu, Antoine-Laurent de. *Rapport de l'un des commissaries charges par le roi de l'éxamen du magnétisme animal*. [1784.] In Alexandre Bertrand, *Du Magnétisme Animal en France*. Paris: J. B. Baillière, 1826.

Kabat-Zinn, Jon. "An Out-Patient Program in Behavioral Medicine for Chronic Pain Patients Based on the Practice of Mindfulness Meditation: Theoretical Considerations and Preliminary Results." *General Hospital Psychiatry* 4 (1982): 33–47.

———. *Full Catastrophe Living: Using the Wisdom of Your Body and Mind to Face Stress, Pain, and Illness*. New York: Delacorte Press, 1990.

Kananjia, Navaz Percy. "*Healing and the Mind*: The Politics of Popular Media and the Development of Mind-Body Medicine in 1990s America." Senior honors thesis, Harvard University, 2001.

Kane, Steven. "Holiness Ritual Fire Handling: Ethnographic and Psychophysiological Considerations." *Ethos* 10, no. 4 (Winter 1982): 369–84.

Kaptchuk, Ted J. "Intentional Ignorance: A History of Blind Assessment and Placebo Controls in Medicine." *Bulletin of the History of Medicine* 72, no. 3 (1998): 389–433.

Katz, Richard. *Boiling Energy: Community Healing among the Kalahari Kung*. Cambridge, Mass.: Harvard University Press, 1982.

Kiecolt-Glaser, Janice K., Ronald Glaser, J. T. Cacioppo, et al. "Marital Conflict in Older Adults: Endocrinological and Immunological Correlates." *Psychosomatic Medicine* 59 (1997): 339–49.

Kiecolt-Glaser, Janice K., Ronald Glaser, S. Gravenstein, et al. "Chronic Stress Alters the Immune Response to Influenza Virus Vaccine in Older Adults." *Proceedings of the National Academy of Sciences* 93 (1996): 3043–47.

Kirin, Narayan. "Refractions of the Field at Home: Hindu Holy Men in America in the Nineteenth and Twentieth Centuries." *Cultural Anthropology* 8 (1993): 476–509.

Kirsch, Irving. *How Expectancies Shape Experience*. Washington, D.C.: American Psychological Association, 1999.

Kleinman, Arthur. *The Illness Narratives: Suffering, Healing, and the Human Condition*. New York: Basic Books, 1988.

Klopfer, Bruno. "Psychological Variables in Human Cancer." *Journal of Projective Techniques* 31, no. 4 (December 1957): 331–40.

Kohn, Livia, and Yoshinobu Sakade, eds. *Taoist Meditation and Longevity Techniques*. Ann Arbor: University of Michigan Center for Chinese Studies, 1989.

Kugelman, Robert. *Stress: The Nature and History of Engineered Grief*. Westport, Conn.: Praeger Publishers, 1992.

Kuriyama, Shigehisa. *The Expressiveness of the Body and the Divergence of Greek and Chinese Medicine*. New York: Zone Books, 1999.

Lachman, Larry. "Group Therapy for all Cancer Patients." *Psychology Today* 35, no. 5 (2002): 27.

Landau, Misia. "Human Evolution as Narrative." *American Scientist* 72 (1984): 262–68.

Lauristen, John. "The Epidemiology of Fear." *New York Native*, August 1, 1988. Reprinted in John Lauristen, *Poison by Prescription: The AZT Story*. New York: Asklepios, 1990. Available online at www.virusmyth.net/aids/data/jlfear.htm.

Lazarus, Richard S. "From Psychological Stress to the Emotions: A History of Changing Outlooks." *Annual Review of Psychology* 44 (1993): 1–21.

Lears, T. J. Jackson. *No Place of Grace: Antimodernism and the Transformation of American Culture, 1880–1920*. New York: Pantheon Books, 1981.

———. "From Salvation to Self-Realization: Advertising and the Therapeutic Roots of the Consumer Culture, 1880–1930." In *The Culture of Consumption: Critical Essays in American History, 1880–1980*, ed. Richard Wightman Fox and T. J. Jackson Lears. New York: Pantheon Books, 1983.

Lee, Mabel, and A. D. Syrokomla-Stefanowska, eds. *Modernization of the Chinese Past*. Broadway, Australia: Wild Peony, 1993.

Lee, Virginia. "Everyday Miracles: An Interview with Dr. Bernie Siegel." Common Ground Online: Resources for Personal Transformation. 1995. Posted online at www.comngrnd.com/siegel.html (this link is no longer active).

Lefferts, Barney. "Chief Guru of the Western World." *New York Times*, December 17, 1967.

Lennon, John, and Jann Wenner. *Lennon Remembers: The Full Rolling Stone Interviews* [from 1970]. New York: Da Capo Press, 2000.

Leonard, B., and K. Miller, eds. *Stress, the Immune System, and Psychiatry*. New York: John Wiley & Sons, 1995.

LeShan, L., and R. E. Worthington. "Personality as a Factor in the Pathogenesis of Cancer." *British Journal of Medical Psychology* 29 (1956): 49–96.

———. "Some Recurrent Life History Patterns Observed in Patients with Malignant Disease." *Journal of Nervous and Mental Diseases* 124 (1956): 460–65.

Lewis, James R., and J. Gordon Melton, eds. *Perspectives on the New Age*. Albany: State University of New York, 1992.

Leys, Ruth. "Traumatic Cures: Shell Shock, Janet, and the Question of Memory." In *Tense Past: Cultural Essays in Trauma and Memory*, ed. Paul Antze and Michael Lambek. New York: Routledge, 1996.

Liébeault, A.-A. *Du sommeil et des états analogues, considérés surtout au point de vue de l'action du morale sur le physique*. Nancy: Nicolas Grosjean, 1866.

Liégeois, J. *De la suggestion et du somnambulisme dans leurs rapports avec la jurisprudence et la médecine légale*. Paris: Octave Doin, 1889.

Linebaugh, Peter. *The London Hanged: Crime and Civil Society in the Eighteenth Century*. London: Verso, 2003.

Lipowski, Z. J. "What Does the Word 'Psychosomatic' Really Mean? A Historical and Semantic Inquiry." *Psychosomatic Medicine* 46, no. 2 (1984): 153–71.

Lippy, Charles, and Peter Williams, eds. *Encyclopedia of the American Religious Experience: Studies of Traditions and Movements*. New York: Scribner, 1988. 3 vol.

Lock, Margaret. *Encounters with Aging: Mythologies of Menopause in Japan and North America*. Berkeley: University of California Press, 1993.

———. "Menopause: Lessons from Anthropology." *Psychosomatic Medicine* 60 (1998): 410–19.

Lopez, Donald S., Jr. *Prisoners of Shangri-La: Tibetan Buddhism and the West*. Chicago: University of Chicago Press, 1998.

Lutz, Antoine, Lawrence L. Greischar, Nancy B. Rawlings, Matthieu Ricard, and Richard

J. Davidson. "Long-term Meditators Self-induce High-amplitude Gamma Synchrony during Mental Practice." *Proceedings of the National Academy of Sciences* 101 (November 16, 2004): 16369–73.

Lutz, Thomas. *American Nervousness, 1903: An Anecdotal History*. Ithaca, N.Y.: Cornell University Press, 1991.

Lynch, James J. *The Broken Heart: The Medical Consequences of Loneliness*. New York: Basic Books, 1977.

———. "Warning: Living Alone Is Dangerous to Your Health." *U.S. News & World Report*, June 30, 1980. Pp. 47–48.

Lyons, John O. *The Invention of the Self: The Hinge of Consciousness in the Eighteenth Century*. Carbondale: Southern Illinois University Press, 1978.

Mackay, Charles. *Extraordinary Popular Delusions and the Madness of Crowds*. [1841.] Foreword by Andrew Tobias. New York: Crown, 1995. Also available online at www.econlib.org/library/mackay/macEx7.html.

Malcolmson, Robert W. *Popular Recreations in English Society, 1700–1850*. New York: Cambridge University Press, 1973.

Marmot, Michael Gideon. "Acculturation and Coronary Heart Disease in Japanese-Americans." Ph.D. diss., University of California, Berkeley, 1975.

———. Social Differentials in Health within and between Populations." In *Health and Wealth*, special volume of *Daedalus* (Proceedings of the American Academy of Arts and Sciences) 123, no. 4 (1994): 197–216.

———. "Redefining Public Health: Epidemiology and Social Stratification." Paper presented for "Conversations with History" at the Institute of International Studies, University of California, Berkeley, March 18, 2002. Available online at http://globetrotter.berkeley.edu/people2/Marmot/marmot-con0.html.

Marmot, Michael Gideon, G. D. Smith, S. Stansfeld, C. Patel, F. North, J. Head, I. White, E. Brunner, and A. Feeney. "Health Inequalities among British Civil Servants: The Whitehall II Study." *Lancet* 337 (1991): 1393–97.

Marmot, Michael Gideon, and S. Leonard Syme. "Acculturation and CHD in Japanese-Americans." *American Journal of Epidemiology* 104 (1976): 225–47.

Marmot, Michael Gideon, and Richard G. Wilkinson, eds. *Social Determinants of Health*. 2d ed. New York: Oxford University Press, 2006.

Mason, John W. "A Historical View of the Stress Field, Part I." *Journal of Human Stress* 1 (March 1975): 10.

Mason, Paul. *The Maharishi: The Biography of the Man Who Gave Transcendental Meditation to the World*. Rockport, Mass.: Element Books, 1994.

Masson, Jeffrey M. *The Assault on Truth: Freud's Suppression of the Seduction Theory*. New York: Farrar, Straus, & Giroux, 1984.

Masuzawa, Tomoko. "From Empire to Utopia: The Effacement of Colonial Markings in *Lost Horizon*." *Positions: East Asia Cultures Critique* 7, no. 2 (1999): 541–72.

Matsumoto, Nancy. "The Burgeoning Art of Healing with the Head: Mind/Body Books Are Big Business." *Los Angeles Times*, September 5, 1994.

Mattingly, Cheryl. *Healing Dramas and Clinical Plots: The Narrative Structure of Experience*. New York: Cambridge University Press, 1998.

McNeill, William H. *Plagues and Peoples*. New York: Doubleday, 1998.

Mechanic, David. "Discussion of Research Programs on Relations between Stressful Life Events and Episodes of Physical Illness." In *Stressful Life Events: Their Nature and*

Effects, ed. Barbara Snell Dohrenwend and Bruce P. Dohrenwend. New York: John Wiley & Sons, 1974.

Mehta, Gita. *Karma Cola: Marketing the Mystic East*. New York: Simon & Schuster, 1979.

Meyer, Donald. *The Positive Thinkers: Religion as Pop Psychology from Mary Eddy to Oral Roberts*. New York: Pantheon Books, 1980.

Micale, Mark S. *Approaching Hysteria: Disease and Its Interpretations*. Princeton, N.J.: Princeton University Press, 1995.

———, and Paul Lerner, eds. *Traumatic Pasts: History, Psychiatry, and Trauma in the Modern Age, 1870–1930*. Cambridge Studies in the History of Medicine. Cambridge, UK: Cambridge University Press, 2002.

Midelfort, H. C. Erik. *Exorcism and Enlightenment: Johann Joseph Gassner and the Demons of Eighteenth-Century Germany*. New Haven, Conn.: Yale University Press, 2005.

Mills, C. Wright. *White Collar: The American Middle Classes*. New York: Oxford University Press, 1951.

Mitchell, Silas Weir. *Fat and Blood, and How to Make Them*. Philadelphia: J. B. Lippincott, 1877.

———. *Doctor and Patient*. Philadelphia: J. B. Lippincott, 1888.

Miura, Kunio. "The Revival of Qi: Qigong in Contemporary China." In *Taoist Meditation and Longevity Techniques*, ed. Livia Kohn and Yoshinobu Sakade. Ann Arbor: University of Michigan Center for Chinese Studies, 1989.

Moerman, Daniel E. *Meaning, Medicine, and the "Placebo Effect."* New York: Cambridge University Press, 2002.

Moyers, Bill, ed. *Healing and the Mind*. New York: Doubleday, 1993.

Murphy, Michael, and Steven Donovan. *The Physical and Psychological Effects of Meditation: A Review of Contemporary Research with a Comprehensive Bibliography, 1931–1996*. Introduction by Eugene Taylor. 2d ed. Sausalito, Calif.: Institute of Noetic Sciences, 1997. Available online at www.noetic.org/research/medbiblio/index.htm.

Myers, C. S. "A Contribution to the Study of Shell Shock: Being an Account of Three Cases of Loss of Memory, Vision, Smell, and Taste, Admitted into the Duchess of Westminster's War Hospital, Le Touquet." *Lancet*, February 13, 1915: 316–20.

Northrup, Christiane. *Women's Bodies, Women's Wisdom: Creating Physical and Emotional Health and Healing*. New York: Bantam Books, 1998.

Olson, Richard. "Spirits, Witches, and Science: Why the Rise of Science Encouraged Belief in the Supernatural in 17th-Century England." *Skeptic* 1, no. 4 (1992): 34–43.

Orme-Johnson, David W., and John T. Farrow, eds. *Scientific Research on Maharishi's Transcendental Meditation and TM-Sidhi Program: Collected Papers*. Vol. 1, 2d ed. West Germany: Maharishi European University Press, 1977.

Ornish, Dean. *Love & Survival: The Scientific Basis for the Healing Power of Intimacy*. New York: HarperCollins, 1998.

Ornish, Dean, S. E. Brown, L. W. Scherwitz, et al. "Lifestyle Changes and Heart Disease." *Lancet* 336 (1990): 741–42.

———. "The Healing Power of Love." *Prevention*, February 1991.

Owen, Richard. "Satan Gets a Facelift." *The Australian*, January 26, 1999. Available online, for a fee, at www.theaustralian.com.au.

Paine, Jeffery. *Re-enchantment: Tibetan Buddhism Comes to the West*. New York: W. W. Norton, 2004.

Parker, Gail Thain. *Mind Cure in New England: From the Civil War to World War I*. Hanover, N.H.: University Press of New England, 1973.

Peale, Norman Vincent. *The Power of Positive Thinking*. New York: Prentice-Hall, 1952.

———. *Stay Alive All Your Life*. Englewood Cliffs, N.J.: Prentice-Hall, 1957.

———. *You Can if You Think You Can*. Englewood Cliffs, N.J.: Prentice-Hall, 1974.

———. *Positive Thinking for a Time Like This*. Englewood Cliffs, N.J.: Prentice-Hall, 1975.

Pepper, O. H. Perry. "A Note on the Placebo." *American Journal of Pharmacy* 117 (1945): 409–12.

———. *Old Doc*. Philadelphia: J. P. Lippincott, 1957.

Perry, Benjamin. "*Qigong*, Daoism, and Science: Some Contexts for the *Qigong* Boom." In *Modernization of the Chinese Past*, ed. Mabel Lee and A. D. Syrokomla-Stefanowska. Broadway, Australia: Wild Peony, 1993.

Pick, Daniel. *Svengali's Web: The Alien Enchanter in Modern Culture*. New Haven, Conn.: Yale University Press, 2000.

Podmore, Frank. *From Mesmer to Christian Science: A Short History of Mental Healing*. New Hyde Park, N.Y.: University Books, 1963.

Porter, Garrett, and Patricia Norris. *Why Me? Harnessing the Healing Power of the Human Spirit*. New York: E. P. Dutton, 1985.

Porter, Roy, and Mikulas Teich, eds. *Romanticism in National Context*. New York: Cambridge University Press, 1988.

Powell, Robert C. "Helen Flanders Dunbar (1902–1959) and a Holistic Approach to Psychosomatic Problems. I. The Rise and Fall of a Medical Philosophy." *Psychiatric Quarterly* 49, no. 2 (1977): 133–52.

———. "Emotionally, Soulfully, Spiritually 'Free to Think and Act': The Helen Flanders Dunbar (1902–59) Memorial Lecture on Psychosomatic Medicine and Pastoral Care." *Journal of Religion and Health* 40, no. 1 (2001): 97–114.

Prince, Raymond, ed. *Trance and Possession States: Proceedings Second Annual Conference, R. M. Bucke Memorial Society, 4–6 March 1966, Montreal*. Montreal: R. M. Bucke Memorial Society, 1968.

Puységur, Marquis de. *Mémoires pour servir à l'histoire et à l'établissement du magnétisme animal*. 2d ed. Paris : Cellot, 1809.

Quimby, P. P. *The Quimby Manuscripts, Showing the Discovery of Spiritual Healing and the Origin of Christian Science*. Ed. Horatio W. Dresser. New York: T. Y. Crowell, 1921. Available online at www.ppquimby.com.

Rapport des Commissaires de la Société Royale de Médecine. [1784.] In Alexandre Bertrand, *Du Magnétisme Animal en France*. Paris: J. B. Baillière, 1826.

Reinhold, Robert. "Exiled Lama Teaches Americans Buddhist Tenets Periled in Tibet." *New York Times*, July 24, 1973.

Reiss, Timothy J. "Denying the Body? Memory and the Dilemmas of History in Descartes." *Journal of the History of Ideas* 57, no. 4 (1996): 587–607.

Reps, Pail, and Nyogen Senzaki, eds. *Zen Flesh, Zen Bones: A Collection of Zen and Pre-Zen Writings*. [1957.] Boston: Charles E. Tuttle, 1998.

Reston, James. "Now Let me Tell You about My Appendectomy in Peking. . . ." *New York Times*, July 26, 1971.

Riska, Elianne. "The Rise and Fall of Type A Man." *Social Science and Medicine* 51 (2000): 1665–74.

———. "From Type A Man to the Hardy Man: Masculinity and Health." *Sociology of Health and Illness* 24 (2001): 347–58.

Rivers, W. H. R. "The Repression of War Experience." *Proceedings of the Royal Society of Medicine* 11 (1918): 1–17.

Rosenberg, Charles E. "The Place of George M. Beard in Nineteenth-Century Psychiatry." *Bulletin of the History of Medicine* 36 (1962): 245–59.

———. "Body and Mind in Nineteenth-Century Medicine: Some Clinical Origins of the Neurosis Construct." *Bulletin of the History of Medicine* 63, no. 2 (1989): 185–97.

———. "Pathologies of Progress: The Idea of Civilization as Risk." *Bulletin of the History of Medicine* 72, no. 4 (1998): 714–30.

Rosenman, Ray H. "Role of Type A Behavior Pattern in the Pathogenesis of Ischemic Heart Disease, and Modification for Prevention." *Advances in Cardiology* 25 (1978): 35–46.

Rosenman, Ray H., and Margaret Chesney. "Type A Behavior Pattern: Its Relationship to Coronary Heart Disease and Its Modification by Behavioral and Pharmacological Approaches." In *Stress in Health and Disease*, ed. Michael R. Zales. New York: Brunner/Mazel, 1985.

Rosenman, Ray H., M. Friedman, R. Straus, et al. "A Predictive Study of Coronary Heart Disease: The Western Collaborative Group Study." *Journal of the American Medical Association* 189 (1964): 15–22.

Roth, Michael. "Hysterical Remembering." *Modernism/Modernity* 3, no. 2 (1996): 1–30.

Rothschild, Babette. *The Body Remembers: The Psychophysiology of Trauma and Trauma Treatment*. New York: W. W. Norton, 2000.

Roush, Wade. "Herbert Benson: Mind-Body Maverick Pushes the Envelope." *Science* New Series 276, no. 5311 (April 18, 1997): 357–59.

Ruberman, William. Review of *The Power of Clan: The Influence of Human Relationships on Heart Disease*, by Stewart Wolf and John G. Bruhn. *New England Journal of Medicine* 329, no. 9 (1993): 669.

Said, Edward. *Orientalism*. New York: Pantheon, 1978.

Satter, Beryl. *Each Mind a Kingdom: American Women, Sexual Purity, and the New Thought Movement, 1875–1920*. Berkeley: University of California Press, 1999.

Scarf, Maggie. "Tuning Down with TM." *New York Times*, February 9, 1975.

Scheper-Hughes, Nancy, and Margaret Lock. "The Mindful Body: A Prolegomenon to Future Work in Medical Anthropology." *Medical Anthropology Quarterly* 1(1987): 6–41.

Schimek, J. G. "Fact and Fantasy in the Seduction Theory: A Historical Review." *Journal of the American Psychoanalytic Association* 35 (1987): 937–65.

Schoenfeld, Amy. "Don't Go Breaking My Heart: A Post–World War II History of Loneliness as a Risk Factor for Cardiovascular Disease." Senior honors thesis, Department of the History of Science, Harvard University, 2007.

Schoepflin, Rennie B. *Christian Science on Trial: Religious Healing in America*. Baltimore: Johns Hopkins University Press, 2003.

Schur, David. "Compulsion as Cure: Contrary Voices in Early Freud." *New Literary History* 32, no. 3 (2001): 585–96.

Selye, Hans. "A Syndrome Produced by Diverse Nocuous Agents." *Nature* 138 (1936): 32.

———. *The Physiology and Pathology of Exposure to Stress: A Treatise Based on the Concepts of the General-Adaptation Syndrome and the Diseases of Adaptation*. Montreal: Acta Incorporated Medical Publishers, 1950.

———. "Implications of the Stress Concept." *New York State Journal of Medicine* 75, no. 12 (October 1975): 2139–44.

———. *The Stress of My Life: A Scientist's Memoirs*. 2d ed. New York: Van Nostrand Reinhold, 1979.

Serinus, Jason, ed. *Psychoimmunity and the Healing Process: A Holistic Approach to Immunity and AIDS*. Berkeley, Calif.: Celestial Arts, 1986.

Shekelle, R. B., R. Gale, A. M. Ostfield, et al. "Hostility, Risk of Coronary Disease, and Mortality." *Psychosomatic Medicine* 45 (1983): 219–28.

Shephard, Ben. *A War of Nerves: Soldiers and Psychiatrists in the Twentieth Century*. Cambridge, Mass.: Harvard University Press, 2001.

Shepherd, Michael. "The Placebo: From Specificity to the Non-specific and Back." *Psychological Medicine* 23 (1993): 1–10.

Shilts, Randy. *And the Band Played On: Politics, People, and the AIDS Epidemic*. New York: St. Martin's Press, 1987.

Shorter, Edward. *From Paralysis to Fatigue: A History of Psychosomatic Illness in the Modern Era*. New York: Free Press, 1992.

Siegel, Bernie. *Love, Medicine, & Miracles: Lessons Learned About Self-Healing from a Surgeon's Experience with Exceptional Patients*. [1986.] New York: HarperPerennial, 1990.

Simonton, O. Carl, Stephanie Matthews-Simonton, and James Creighton. *Getting Well Again: A Step-by-Step Self-Help Guide to Overcoming Cancer for Patients and Their Families*. Los Angeles: J. P. Tarcher, 1978.

Smelser, Neil J., and Paul B. Baltes, editors in chief. *International Encyclopedia of the Social and Behavioral Sciences*. New York: Elsevier, 2001.

Solomon, George, and Lydia Temoshok. "A Psychoneuroimmunologic Perspective on AIDS." In *Psychosocial Perspectives on AIDS: Etiology, Prevention, and Treatment*, ed. Lydia Temoshok and Andrew Baum. Hillsdale, N.J.: Lawrence Erlbaum, 1990.

Solomon, George F., and R. H. Moos. "The Relationship of Personality to the Presence of Rheumatoid Factor in Asymptomatic Relatives of Patients with Rheumatoid Arthritis." *Psychosomatic Medicine* 27 (1965): 350–60.

Sontag, Susan. *Illness as Metaphor, and AIDS and Its Metaphors*. New York: Picador, 2001. *Illness as Metaphor* originally published in 1978.

Spanos, Nicholas, and Jack Gottleib. "Demonic Possession, Mesmerism, and Hysteria: A Social Psychological Perspective on Their Historical Interrelation." *Journal of Abnormal Psychology* 88 (1979): 527–46.

Spiegel, David. "Social Support: How Friends, Family, and Groups Can Help." In *Mind, Body Medicine: How to Use Your Mind for Better Health*, ed. Daniel Goleman and Joel Gurin. Yonkers, N.Y.: Consumer Reports Books, 1993.

———. *Living Beyond Limits: New Hope and Help for Facing Life-Threatening Illness*. New York: Random House, 1994.

Spiegel, David, J. R. Bloom, H. C. Kraemer, and E. Gottheil. "Effect of Psychosocial Treatment on Survival of Patients with Metastatic Breast Cancer." *Lancet* 2, no. 8668 (1989): 888–91.

Spiegel, David, J. R. Bloom, and Irvin D. Yalom. "Group Support for Patients with Metastatic Cancer: A Randomized Prospective Outcome Study." *Archives of General Psychiatry* 38 (1981): 527–33.

Spiegel, David, Lisa D. Butler, Janine Giese-Davis, Cheryl Koopman, Elaine Miller, Sue DiMiceli, Catherine Classen, Patricia Fobair, Robert W. Carlson, Helena Kraemer. "Effects of Supportive-Expressive Group Therapy on Survival of Patients with Metastatic Breast Cancer: A Randomized Prospective Trial." *Cancer* (July 23, 2007). Available online at http://dx.doi.org/10.1002/cncr.22890.

Spiro, Howard. *The Power of Hope: A Doctor's Perspective.* New Haven, Conn.: Yale University Press, 1998.

Spitz, R. A. "Hospitalism: An Inquiry into the Genesis of Psychiatric Conditions in Early Childhood, Part I." *Psychoanalytic Studies of the Child* 1 (1945): 53–74.

———. "Hospitalism: An Inquiry into the Genesis of Psychiatric Conditions in Early Childhood, Part II." *Psychoanalytic Studies of the Child* 2 (1945): 113–17.

Stark, Rodney. "The Rise and Fall of Christian Science." *Journal of Contemporary Religion* 13, no. 2 (1998): 189–214.

Stepansky, P. E., ed. *Freud: Appraisals and Reappraisals—Contributions to Freud Studies.* Vol. 1. Hillsdale, N.J.: Analytic Press, 1986.

Stout, C., J. Morrow, E. N. Brandt, Jr., and S. Wolf. "Unusually Low Incidence of Death from Myocardial Infarction: Study of an Italian-American Community in Pennsylvania." *Journal of the American Medical Association* 188 (1964): 845–49.

Streitfeld, Richard. "Mindful Medicine: An Interview with Jon Kabat-Zinn." *Primary Point* 8, no. 2 (Summer 1991). Available online at www.kwanumzen.com/primarypoint/.

Syme, S. Leonard. "Historical Perspective: The Social Determinants of Disease—Some Roots of the Movement." *Epidemiologic Perspectives & Innovations* 2, no. 2 (2005). Available online at www.epi-perspectives.com/content/2/1/2.

Szabo, Jason. "Seeing Is Believing? The Form and Substance of French Medical Debates over Lourdes." *Bulletin of the History of Medicine* 76, no. 2 (2002): 199–230.

Szabo, Sandor. "Hans Selye and the Development of the Stress Concept: Special Reference to Gastroduodenal Ulcerogenesis." *Annals of the New York Academy of Sciences* 851 (1998): 19–27.

Taylor, Eugene. *Shadow Culture: Psychology and Spirituality in America.* Washington, D.C.: Counterpoint, 1999.

———. "A Perfect Correlation Between Mind and Brain: William James's Varieties and the Contemporary Field of Mind/Body Medicine." *The Journal of Speculative Philosophy* 17, no. 1 (2003): 40–52.

Taylor, Shelley E. *The Tending Instinct: How Nurturing is Essential for Who We Are and How We Live*. New York: Owl Books, 2002.

Temoshok, Lydia. "Biopsychosocial Studies on Cutaneous Malignant Melanoma: Psychosocial Factors Associated with Prognostic Indicators, Progression, Psychophysiology, and Tumor-Host Response." *Social Science and Medicine* 20 (1985): 833–40.

Temoshok, Lydia, and Andrew Baum, eds. *Psychosocial Perspectives on AIDS: Etiology, Prevention, and Treatment*. Hillsdale, N.J: Lawrence Erlbaum, 1990.

Temoshok, Lydia, and Henry Dreher. *The Type C Connection: The Behavioral Links to Cancer and Your Health*. New York: Random House, 1992.

Thomas, Kathy Quinn. "The Mind-Body Connection: Granny Was Right, After All."

Rochester Review 59, no. 3 (1997). Available online at www.rochester.edu/pr/Review/V59N3/feature2.html.

Thomas, Keith. *Religion and the Decline of Magic: Studies in Popular Beliefs in Sixteenth and Seventeenth Century England*. London: Weidenfeld & Nicolson, 1971.

Toffler, Alvin. *Future Shock*. New York: Random House, 1970.

Toshim, H., Y. Koga, and H. Blackburn, eds. *Lessons for Science from the Seven Countries Study: A 35-year Collaborative Experience in Cardiovascular Disease Epidemiology*. New York: Springer Verlag, 1994.

Towler, Solala. "QiGong Comes West—From Chronic Fatigue to Vital Energy." *Alternatives: Resources for Cultural Creativity* 14 (Summer 2000). Available online at www.alternativesmagazine.com/14/towler.html.

Trine, Ralph Waldo. *In Tune with the Infinite or, Fullness of Peace, Power, and Plenty*. New York: Dodge Publishing, 1919.

Turner, B. I. "Social Capital, Inequality, and Health: The Durkheimian Revival." *Social Theory & Health* 1, no. 1 (May 2003): 4–20.

Van der Kolk, Bessel A. "The Body Keeps the Score: Approaches to the Psychobiology of Posttraumatic Stress Disorder." In *Traumatic Stress: The Effects of Overwhelming Experience on Mind, Body, and Society*, ed. Bessel A. Van der Kolk, A. C. McFarlane, and L. Weisaeth. New York: Guilford, 1996.

Van der Kolk, Bessel A., and R. Fisler. "Dissociation and the Fragmentary Nature of Traumatic Memories: Overview and Exploratory Study." *Journal of Traumatic Stress* 8 (1995): 505–25.

Van der Kolk, Bessel A., A. C. McFarlane, and L. Weisaeth, eds. *Traumatic Stress: The Effects of Overwhelming Experience on Mind, Body, and Society*. New York: Guilford, 1996.

Vecsey, George. "Dalai Lama, in Texas, Hears Debate among Learned Men." *New York Times*, September 20, 1979.

Vickers, Andrew. "Against Mind-Body Medicine." *Complementary Therapies in Medicine* 6 (1998): 111–14.

Vila, Anne C. "The Making of the Modern Body." *Modern Language Notes* 104, no. 4 (September 1989): 927–36.

Viner, Russell. "Putting Stress in Life: Hans Selye and the Making of Stress Theory." *Social Studies of Science* 29, no. 3 (1999): 391–410.

Walker, D. P. *Unclean Spirits: Possession and Exorcism in France and England in the Late Sixteenth and Early Seventeenth Centuries*. London: Scolar Press, 1981.

Wallace, Alan B., ed. *Buddhism and Science: Breaking New Ground*. New York: Columbia University Press, 2003.

Wallace, Robert Keith. "Physiological Effects of Transcendental Meditation." *Science* 167 (1970): 1751–54.

———. "Proposed Fourth Major State of Consciousness." In *Scientific Research on Maharishi's Transcendental Meditation and TM-Sidhi Program: Collected Papers*, vol. 1, 2d ed., ed. David W. Orme-Johnson and John T. Farrow. West Germany: Maharishi European University Press, 1977. Pp. 43–78.

Wallace, Robert Keith, Herbert Benson, and Archie F. Wilson. "A Wakeful Hypometabolic Physiologic State." *American Journal of Physiology* 221, no. 3 (September 1971): 795–99.

Watts, Alan. *Behold the Spirit: A Study of the Necessity of Mystical Religion*. [1947.] New York: Pantheon Books, 1971.

Weizsäcker, Viktor von. "Über medizinische Anthropolgie." [1927.] In *Gesammelte Schriften in 10 Bänden, Band 5: Der Arzt und der Kranke: Stücke einer medizinischen Anthropologie*, ed. Peter Achilles, Dieter Janz, Martin Schrenk, and Carl Friedrich von Weizsäcker. Frankfurt am Main: Surkamp Verlag, 1986.

———. "Der Arzt und der Kranke: Stücke einer medizinischen Anthropologie." *Die Kreatur* 1 (1927): 69–86.

———. "Studien zur Pathogenese: Angina Tonsillaris." [1946.] In *Psychosomatische Medizin: Wege der Forschung*, ed. Peter Hahn. Darmstadt: Wissenschaftliche Buchgesellschaft, 1985.

———. *Gesammelte Schriften*. 9 vol. Ed. Peter Achilles, Dieter Janz, Martin Schrenk, and Carl Friedrich von Weizsäcker. Frankfurt am Main: Surkamp Verlag, 1986–1987.

Weldon, John, and Paul Carden. "Ernest Holmes and Religious Science." *The News and Research Periodical of the Christian Research Institute* 7, no. 1 (1984), available online at http://associate.com/ministry_files/The_Reading_Room/False_Teaching_n_Teachers_3/Religious_Science.shtml.

Whitehouse, Deb. "Review of *Mary Baker Eddy* by Gillian Gill." *Journal of the Society for the Study of Metaphysical Religion* 5, no. 1 (Spring 1999): 75–79. Available online at http://websyte.com/alan/eddyrevw.htm.

Whorton, James. *Nature Cures: The History of Alternative Medicine in America*. New York: Oxford University Press, 2004.

Whyte, William. *The Organization Man*. New York: Simon & Schuster, 1956.

Widdowson, E. M. "Mental Contentment and Physical Growth." *Lancet* 1 (1951): 1316–18.

Will, Barbara. "The Nervous Origins of the American Western." *American Literature* 70, no. 2 (1998): 293–316.

Williams, J. M., R. G. Peterson, P. A. Shea, J. F. Schmedtje, D. C. Bauer, and D. L. Felten. "Sympathetic Innervation of Murine Thymus and Spleen: Evidence for a Functional Link between the Nervous and Immune Systems." *Brain Research Bulletin* 6 (1981): 83–94.

Williams, Redford. *The Trusting Heart: Great News about Type A Behavior*. New York: Times Books, 1989.

Winkelman, Michael. *Shamanism: The Neural Ecology of Consciousness and Healing*. Westport, Conn.: Bergin & Garvey, 2000.

Winter, Alison. "Mesmerism and the Introduction of Surgical Anesthesia to Victorian England." *Engineering & Science* 2 (1998): 20–37.

———. *Mesmerized: Powers of Mind in Victorian England*. Chicago: University of Chicago Press, 1998.

Wolf, Stewart, and John G. Bruhn. *The Power of Clan: The Influence of Human Relationships on Heart Disease*. New Brunswick, N.J.: Transaction Publishers, 1993.

Wolf, Stewart, and Harold Wolf. *Human Gastric Function: An Experimental Study of a Man and His Stomach*, with a foreword by Walter B. Cannon. New York: Oxford University Press, 1943.

Wolfe, Elin L., A. Clifford Barger, and Saul Benison. *Walter B. Cannon: Science and Society*. Cambridge, Mass.: Harvard University Press, 2000.

Wright, Peggy Ann. "The Nature of the Shamanistic State of Consciousness: A Review." *Journal of Psychoactive Drugs* 21, no. 1 (January–March 1989): 25–33.

Xu, Guagqiu. "The United States and the Tibet Issue." *Asian Survey* 37, no. 11 (November 1997): 1062–77.

Xu, Jian. "Body, Discourse, and the Cultural Politics of Contemporary Chinese Qigong." *Journal of Asian Studies* 58 (November 1999): 961–91.

Yalom, Irvin. *Existential Psychotherapy.* New York: Basic Books, 1980.

Yogi, Maharishi Mahesh. *Meditations of Maharishi Mahesh Yogi*. New York: Bantam Books, 1973.

"You Changed America's Heart: 1948–1998: A 50th Anniversary Tribute to the Participants in the Framingham Heart Study." Available online at www.nhlbi.nih.gov/about/framingham/fhsbro.htm.

Young, Allan. *The Harmony of Illusions: Inventing Posttraumatic Stress Disorder,* Princeton, N.J.: Princeton University Press, 1995.

Young, J. H. *Medical Messiahs: A Social History of Health Quackery in Twentieth-Century America*. Princeton, N.J.: Princeton University Press, 1967.

Zales, Michael R., ed. *Stress in Health and Disease*. New York: Brunner/Mazel, 1985.

INDEX

Page numbers in *italics* refer to illustrations.